SpringerBriefs in Earth Sciences

For further volumes:
http://www.springer.com/series/8897

Rituparna Bose

Palaeobiology of Middle Paleozoic Marine Brachiopods

A Case Study of Extinct Organisms in Classical Paleontology

 Springer

Rituparna Bose
City University of New York
New York, NY
USA

ISSN 2191-5369   ISSN 2191-5377   (electronic)
ISBN 978-3-319-00193-7   ISBN 978-3-319-00194-4   (eBook)
DOI 10.1007/978-3-319-00194-4
Springer Cham Heidelberg New York Dordrecht London

Library of Congress Control Number: 2013935487

Printed on acid-free paper

Springer is part of Springer Science+Business Media (www.springer.com)

Foreword

Brachiopods are among the most ubiquitous of all Paleozoic groups, and their relatively high preservation potential has resulted in an exceptionally rich fossil record. Although there are approximately 100 extant genera, perhaps 50 times this number are known only from fossils. Brachiopods therefore represent an extant clade that was once vastly more diverse, containing species that were a much more staple and important component of marine ecosystems than they are today. As a group, their diversity was little more than dented by the End-Ordovician, Frasnian, and Serpukhovian events, and they maintained their dominance of most infaunal communities throughout the Paleozoic. Only the End-Permian event severely compromised their diversity, with brachiopods giving way to bivalves as the dominant infaunal clade during the subsequent post-Paleozoic recovery. This book considers their mid-Paleozoic heyday.

The fossil record may offer valuable insights into the current biodiversity crisis. It records 'natural experiments' in which groups have repeatedly faced environmental challenges or habitat destruction, and may therefore document patterns of extinction susceptibility and resistance. It also records in which groups have been able to reradiate and diversify in the wake of these environmental upheavals. Thorough scrutiny and analysis of fossil data may therefore yield general principles that can be used to predict the probable responses of living species in the current phase of extinction. Underpinning all such inferences is taxonomy of a uniformly high standard, coupled with detailed stratigraphic, paleobiogeographical, morphological, and ecological data. Dr. Bose makes very significant strides in this direction here.

This book offers an excellent introduction to the paleobiology of mid-Paleozoic brachiopods. It also contains a rich collation of landmark data describing their shape, stratigraphic data describing their temporal ranges, and paleogeographic data detailing their spatial distributions. In combination, these resources enable several novel analyses, including plots of diversity and taxonomic turnover through time, analyses of morphological disparity, and clade dispersal through empirical morphospaces, as well as investigations of speciation mode. In addition, the author discusses the role of brachiopods in extinct ecosystems, as well as the manner in which these ecosystems changed during periods of increased turnover. The book

therefore offers an invaluable reference for graduate students and others seeking
to make macroevolutionary and macroecological inferences from paleobiological
data. More specifically, it offers an excellent complement to the usual resources on
brachiopod taxonomy and evolution.

Dr. Matthew A. Wills
Reader in Evolutionary Biology
Department of Biology and Biochemistry
University of Bath

Preface

The prerequisite to developing effective strategies for conserving biodiversity is a profound understanding of the taxonomy, evolution, and ecology of all life forms. It is especially important to comprehend the link between evolution, ecology, and environment and perhaps, appreciate the significance of such studies in extinct organisms; especially in organisms that were abundant in a certain geologic era, but have subsequently dwindled or become extinct. Such studies should help to understand extinction, accurately gauge the underlying causes behind loss of biodiversity and make predictions about future distribution of biodiversity. I apply novel quantitative techniques to track biodiversity loss, what should also serve as a starting point for conservation.

An increasing number of species are becoming extinct at an alarming rate today. This will soon lead to a colossal biodiversity crisis; and eventually to the paucity of non-renewable resources of energy making our Earth unsustainable in future. To save our mother planet from this crisis, studies need to be performed at large to discover abundant new fossil sites on Earth for continued access to oil-rich locations. Most importantly, a holistic approach is necessary in solving the present problem of biodiversity loss. This book presents the use of advanced quantitative models in understanding emerging topics in evolutionary biology that include biodiversity, taxonomy, phylogeny, evolution, and ecology of extinct organisms.

Traditionally, the broader view was that ecological interactions occurred in such short time scales than evolution could be easily ignored. A recent study by an evolutionary biologist, Dr. David Reznick in University of California Riverside has shown that certain organisms (freshwater fish) can evolve rapidly in response to ecological interactions. Thus, it is ecology that shapes evolution. These significant results are now published in the journal, Proceedings of the National Academy of Sciences. My study on brachiopods is unique in that it involves quantification of the ecological consequences of both slow and rapid adaptation of organisms, which is also known as the evolutionary response of organisms to environment. Ecology also has a direct association with biodiversity; with changing ecological conditions, biodiversity of organisms can also change. Thus, this book will assist future evolutionary biologists in understanding the natural and anthropogenic causes behind biodiversity crisis and ecosystem collapse. Besides evolutionary biologists, paleoecologists, evolutionary ecologists, systematists, paleoclimatologists, and conservationists, this study would be of great interest to explore geologists and geophysicists in potentially unraveling natural resources from our sustainable Earth.

The author is grateful to Prof. David Polly (Associate Professor of Geological Sciences in the Indiana University) for his valuable suggestions.

Acknowledgments

A special thanks to Carlton Brett, Ed Landing, Neil Landman, Susan Butts, Donald Hattin, Erle Kauffman, Abhijit Basu, Claudia Johnson and David Polly.

I would also like to acknowledge Murat Maga for providing illustrations on brachiopod internal morphology. Finally, my deepest appreciation for Arthur Boucot for providing a careful review of the manuscript.

Financial support for this research was derived primarily from Galloway-Horowitz Research Grant-in-Aid granted by Department of Geological Sciences, Indiana University; Indiana University School of Arts and Sciences Dissertation Year Research Fellowship and Theodore Roosevelt Memorial Grant, American Museum of Natural History.

Contents

Abstract

Fossil species appear to persist morphologically unchanged for long intervals of geologic time, punctuated by short bursts of rapid change as explained by the Ecological Evolutionary Units (EEUs). Here, we investigated morphological variation in Paleozoic atrypide morphology at the subfamily level from the Silurian and Devonian time intervals in the third Paleozoic EEU (~444-359 my) using relatively new techniques of quantitative modeling. We measured valve shape in 1550 atrypide individuals from 6 large-scale time intervals (Atrypinae: Early Silurian–Early Devonian; Variatrypinae: Middle Devonian–Late Devonian) from 7 EE subunits (Clinton, Lockport, Keyser, Heidelberg, Schoharie, Onondaga, and Hamilton) from Eastern North America region, 1193 of which were used to assess geographic variation within 3 time intervals (Middle Silurian = 329, Early Devonian = 406, and Middle Devonian = 458).

Atrypinae and Variatrypinae have similar spiralia, cardinal processes, and jugal processes. Atrypinae appeared in Llandovery and persisted through Emsian (Early Devonian) while Variatrypinae appeared in the Pragian and persisted through Frasnian. Using these data, 4 hypotheses were tested: (1) If these 2 subfamilies are closely related to each other, then we would expect less significant differences between samples from Early Silurian–Early Devonian (Atrypinae dominated) and Middle Devonian–Late Devonian (Variatrypinae dominated) time intervals. MANOVA showed significant shape differences between different time horizons ($p \leq 0.01$), suggesting some morphological variation between the two subfamilies; (2) If subfamily Variatrypinae was derived from the Atrypinae subfamily, and atrypide individuals in these subfamilies evolved in a gradual, directional manner, then we would expect samples from Early Silurian–Early Devonian and Middle Devonian–Late Devonian form two separate clusters with their groups being closely linked to each other. Euclidean-based cluster analysis shows samples widely separated in time are more similar with no grouping of time units within two subfamilies; (3) If these 2 subfamilies are cosmopolitan, then we would expect significant difference in atrypide morphological shape within Atrypinae subfamily analyzed from the Middle Silurian through Early Devonian geographic localities and Variatrypinae subfamily analyzed from Middle Devonian geographic localities; Multivariate analyses shows that there is significant difference in morphological shape in these subfamilies from each time interval ($p \leq 0.01$) further suggesting some geographic variation; (4) If Atrypinae and Variatrypinae subfamilies are closely linked and the later derived from the former, then their morphological patterns must be closely similar. If some morphological variation occurs, it could be due to the ecological interactions between host atrypides and other episkeletobionts. Statistical correlation between morphological change in valves and encrustation rate from each time interval is relatively significant suggesting ecological stasis might be driving the morphological patterns observed.

Atrypide individuals from both subfamilies show close to similar mean morphological shape and similar morphological trends with clear overlap between the lowermost and uppermost occurrences. This is an indicator of morphological conservation in these subfamilies further implying stasis-like patterns in the atrypide group. Moderate geographic variation in atrypide individuals during the middle Silurian, early and middle Devonian time from Eastern North America region suggest this differentiation is likely due to the speciation events in Atrypinae and Variatrypinae subfamilies. Morphological shape change patterns observed in atrypides could be due to ecological interactions between hosts and episkeletobionts. Lithological variations and sea level changes during the Silurian and Devonian period result in stable morphologies which supports Sheldon's Plus ca model. Thus, this study explains how a group of closely related taxa in atrypide subfamilies exhibit morphological conservation through time in P3 EEU within the Eastern North America region.

Chapter 1
Introduction

1.1 Ecological Evolutionary Subunits

Large scale periods of stability in fossil communities are known as Ecological Evolutionary (EEUs) Units (Boucot 1983, 1986, 1990; Sheehan 1991, 1996; Holterhoff 1996), and they are separated by periods of rapid reorganization in community interrelations (Brett et al. 1990; Brett and Baird 1995; Holterhoff 1996; Ivany et al. 2009). In other words, little substantial change occurs in relative abundance and diversity of communities within these EEUs (Brett and Baird 1995) with morphology of component taxa remaining the same (Ivany et al. 2009) while major extinction events and evolutionary radiations mark the EEU boundaries. This concept of EEU (Boucot 1983) was defined based on communities. The accompanying stability of species morphologies within these EEUs is due to "ecological locking," a phenomena operated by natural selection (Morris 1995; Morris et al. 1995). The 12 EEUs in the Phanerozoic (Boucot 1983) were revised to 9 EEUs after eliminating recovery intervals and given that, Silurian-Devonian time represents a single EEU (EEU P3) marked by periods of stability (several EE subunits) interspersed periodically by periods of reorganization and extinction. Major extinction events occurred in the Early Silurian and Late Devonian time that marked the P3 EEU boundary (Sheehan 1996, Fig. 3). The EE subunits that have been studied from the P3 EEU are listed below in Fig. 1.1.

The paleontological record of the lower and middle Paleozoic Appalachian foreland basin demonstrates an unprecedented level of ecological and morphological stability on geological time scales. Some 70–80 % of fossil morphospecies within assemblages persisted in similar relative abundances in coordinated packages lasting as long as 7 million years despite evidence for environmental change and biotic disturbances (Morris et al. 1995). This phenomenal evolutionary stability despite environmental fluctuations has been explained by the concept of ecological locking. Ecological locking provides the source of evolutionary stability that is suggested to have been caused by ecological interactions that maintain a static adaptive landscape and prevent both the long-term establishment of exotic invading species and

R. Bose, *Palaeobiology of Middle Paleozoic Marine Brachiopods*,
SpringerBriefs in Earth Sciences, DOI: 10.1007/978-3-319-00194-4_1,
© The Author(s) 2013

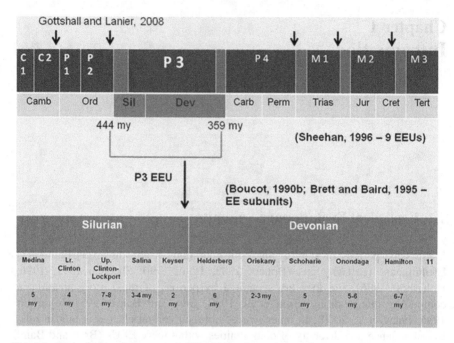

Fig. 1.1 Ecological evolutionary unit P3 showing the major subdivided 11 EE subunits in the Silurian and Devonian

evolutionary change of native species (Morris et al. 1995). Only when disturbance exceeds the capacity of the ecosystem do ecological crashes occur, and evolution proceeds at high rates of directional selection during the organization of a new stable ecological hierarchy (Morris et al. 1995). Thus these EEUs have been used to explain why many fossil species appear to persist unchanged morphologically for long periods, punctuated by short bursts of rapid change. This study investigates the morphological shape change pattern in atrypide brachiopods at the subfamily level both in time and space in an evolutionary perspective. This study more specifically is designed to determine comparative morphological shape patterns within the atrypide brachiopod individuals of the Atrypinae and Variatrypinae subfamilies from the Silurian and Devonian EE subunits (P3 EEU; Sheehan 1996) from Appalachian Basin strata and other time equivalent strata (EE subunits—Clinton, Lockport, Helderberg, Oriskany, Schoharie, Onondaga, Hamilton, subunit 11) that span the entire 64 million year of the Silurian-Devonian (441–376 million year) rock record (Gradstein et al. 2005) (Fig. 1.1).

1.2 Patterns of Morphological Change in Fossil Lineages

Species in the fossil record are expected by some to be geographically homogenous entities that persist roughly unchanged through their history, with most morphological change occurring at speciation events (Elredge and Gould 1972; Gould and Elredge 1977). The relative importance of stasis has been studied for many fossil lineages. Ecological Evolutionary subunits are short term intervals (3–7 million year) of community stability within the larger Ecological Evolutionary units and within a typical Ecological Evolutionary subunit, morphospecies appear to persist with little or no morphological change (Brett and Baird 1995). Shape traits like bivalve convexity studied for Neogene *Chesapecten nefrens* for over 4 million year have yielded results in support for stasis (Hunt 2007). Several other fossil lineages have been put together in a data set by Hunt (2007) who found that stasis is supported by 50 % of the evolutionary models as opposed to directional evolution.

Brachiopod fossil lineages have been studied in the past within Paleozoic EE subunits. Lieberman et al. (1995) observed morphological overlap within 2 brachiopod species lineages (*Mediospirifer audaculus* and *Athyris spiriferoides*) between the lowermost and uppermost strata with some variations in the intervening samples of the Hamilton Group (Hamilton EE subunit). and Goldman and Mitchell (1990) tested the internal morphology of three brachiopod species of the Hamilton Group of western New York from size measurements and found only one species of Late Givetian age showed some species level change within the Hamilton EE subunit. Isaacson and Perry (1977) have not found any significant change in *Tropidoleptus carinatus* of the Givetian age Hamilton Group from the lowest to its highest occurrence, spanning some 40 million year and further, Elrdege tested the same fauna using morphometrics and found almost no significant morphological change in this unit. Overall, a majority of Hamilton brachiopod species lineages represent stasis in most cases while some minor evolutionary changes have been recorded in some cases. Unidirectional evolution over geological time intervals is highly unlikely and that evolutionary reversals are common (Sheldon 1996). Haney et al. (2001) in a different study found that brachiopod valve shape in Devonian EE subunits did not change, but that brachiopods in Ordovician EE subunits changed more than expected from this paradigm. So does brachiopod morphological shape normally conform to the EE hypothesis of stasis? In my study, I will test whether morphological stasis or change is more common in brachiopod shape by looking at Silurian and Devonian atrypides from successive EE subunits within a single EEU. Thus, it is important to determine if brachiopod morphological shape normally conforms to the EE hypothesis of stasis.

Not all workers agree that stasis is common within fossil lineages. Two studies that have most prominently challenged stasis are Gingerich's (1976) work on fossil mammals from the Bighorn Basin of the western United States and Sheldon's (1987, Fig. 6) work on Ordovician trilobites from Wales. Sheldon (1987) reported evidence of phyletic gradualism based on a study of eight lineages of 15000 Ordovician trilobites from central Wales over a 3 million year interval. He believed that subdividing a species lineage into subspecies often gives a false impression of punctuation and stasis. However, in a later study, Sheldon (1996) proposed the Plus ~ca change model that predicts a tendency for continuous, gradualistic evolution in the tropical zones and in the deep sea (narrowly fluctuating, relatively stable environments), and more stasis with episodic punctuations in shallow waters and temperate zones (widely fluctuating environments with its application in more physical environmental variables like sea level, substrate, temperature, etc. on a geologic time scale) (Sheldon 1996, Fig. 2; Sheldon 1990). Webber and Hunda (2007) using geometric morphometrics found that certain aspects of the morphological shape of the Upper Ordovician trilobite *Flexicalymene granulosa* changes with change in paleoenvironmental conditions during the deposition of Kope and lower Fairview Formations spanning over 2 million year interval. We will study whether morphological shape changes or remains the same over time and space for a longer time interval 64 million year (441–376 million year).

Studying shape patterns in brachiopod individuals from different geographic localities would help investigating the species morphological distribution in the ENA region. Haney and Mitchell (1998) studied the geographic variation in Ordovician brachiopod *Sowerbyella curdsvillensis* from Virginia, Kentucky, New York, and Quebec using landmark based morphometric technique. We have investigated the morphological variation in atrypide group from different geographic localities during the time intervals (Middle Silurian, Early and Middle Devonian) chosen for investigation which will help future researchers in determining the species distribution within this group in this region.

The main significance of this study is in testing the morphological shape variability of the atrypide brachiopod individuals at the subfamily level within a large time scale in P3 EEU to see how a group of closely related taxa (Subfamilies Atrypinae and Variatrypinae) have changed morphologically through time within one region. Besides, brachiopod morphological shapes will be determined from various geographic locations at the subfamily level to determine their shape variation in space.

1.3 Why Atrypides

For this study, atrypides were selected as these are the oldest brachiopods to have a mineralized lophophore support called the spiralium to support their complex (coiled) lophophore structure (Rudwick 1965a, 1970; Carlson and Leighton 2001). The evolutionary trends of the spiralia and jugum in primitive atrypide brachiopods have been studied by Copper (1977). The morphological shape of the brachiopod reflects the spiralia/lophophore shape (Rudwick 1970) and recently quantitative studies have been performed to detect species differentiation in atrypides based on their morphological shape (Bose 2012a, b, c, Bose and De 2013). Thus, studying their morphological shape is important in paleontological studies. A quick illustration of the internal structure of *Desquamatia* sp. (ventral valve) from the subfamily Variatrypinae is included in this study to show their brachidium support for the spirolophe (Fig. 3). More studies have been performed on internal morphological characters of brachial and pedicle valves (Bowen 1966; Copper 1967).

Filtration capacity in brachiopods is a function of the area of the rows of the filaments and thus of the total surface area of the cones formed by the spiralia (Rudwick 1958, 1960, 1970). Rudwick (1970, Fig. 82), argued that *Atrypa* and *Spirifer* had lateral inhalant currents and a medial exhalant current and he based his conclusion on a comparison with modern brachiopods and in particular

Crania. As brachiopods grow, their metabolic needs increase roughly in proportion to the volume of the body and this demand is met by an increase in the size of the lophophore (Hancock 1858). Campbell and Chatterton (1979) and Rudwick (1960) noted that the spiralia of fossil spiriferids are quite closely 'moulded' to the shape of the body-cavity, and this certainly appears to be the case for Atrypoidea (Pl. 1, Fig. 2). It can be argued therefore, that the increase in size and convexity of Atrypoidea and Protathyris during ontogeny (Jones 1974, 1977, 1982) may be a measure of the increase in size of the spiralia and hence reflects the animals need to increase its filtering capacity during ontogeny. Thus, studying the brachiopod morphological shape in these two atrypide subfamilies could provide evidence of their lophophore and spiralia structure during life.

The taxonomy of the atrypide brachiopod group, previously taxonomized as *Atrypa reticularis*, a single collective species in America (Fenton and Fenton 1930) has been revised to the genus level based on recent literature published (Copper 1973, 1996, 2001, 2002, 2004; Day 1998; Day and Copper 1998). *Atrypa reticularis* is sensostricto from the Silurian of Gotland (Jed Day and Paul Copper pers. comm.). Atrypides increased in generic diversity and abundance during early Silurian (Upper Llandovery to Wenlockian) with decline in diversity during upper Silurian (Pridoli) that persisted through early Devonian (Lockhovian) followed by a peak in diversity during Emsian-Givetian when many of these genera co-existed worldwide (Copper 2001, Fig. 21.1). Generic diversity identified in Atrypinae and Variatrypinae subfamilies based on external morphological characters in the Silurian include the taxa: *Atrypa, Gotatrypa, Endrea, Joviatrypa, Protatrypa, Plectatrypa* and those from the Devonian include *Atrypa, Kyrtatrypa, Spinatrypa, Desquamatia (Desquamatia), Desquamatia (Independatrypa),* and *Pseudoatrypa* (Copper 2001, 2004; R. Bose personal observation) with taxa from Atrypinae subfamily dominating our Early Silurian–Early Devonian samples and taxa from Variatrypinae subfamily dominating our Middle–Late Devonian samples. A phylogenetic analysis based on internal morphological characters of these diverse atrypide genera in Atrypinae and Variatrypinae subfamilies from the Eastern North American region could resolve the pending issue of the relatedness of these taxa in these 2 subfamilies which is reserved for our future analysis.

Atrypide individuals that have been measured for this study, belong to closely related subfamilies—Atrypinae and Variatrypinae which have similar dorsally to dorso-medially directed (Copper 2002) spiralia, cardinal processes and jugal processes. Atrypinae adults usually lack the apical foramen as opposed to Variatrypinae. Atrypinae individuals have wavelike-very wide overlapping or imbricate growth lamellae extended as frills while Variatrypiane individuals have widely spaced growth lamellae extended as expansive frills (Copper 2002). Atrypinae appeared during Lower Silurian (Llandovery) and dispappeared in late Early Devonian (Emsian?) while Variatrypinae appeared in Lower Devonian (Pragian) and disappeared during the Late Devonian (Frasnian) (Copper 1997, 2002). My data from the Lower Silurian to Upper Silurian time units consist of atrypide individuals that belong to the Atrypinae subfamily while the Middle Devonian to Upper Devonian time units consist of those from the Variatrypiane subfamily.

1.4 Climate and Environment in the Silurian and Devonian

Silurian and Devonian periods were times of widespread reef development, carbonate deposition, and orogenic activities (Boucot, 1981; Stanley 1999) with fauna inhabiting mostly subtropical climate zones with some extensions to temperate zones. Eastern North America was transformed from an Early Silurian highland to a Middle Silurian carbonate shelf, and reefs and evaporite deposition continued in Western North America. Appalachian mountains were formed when Laurentia, Baltica and Avalonia (part of Gondwanaland) collided to form Euramerica. Acadian orogeny played a major role from middle Silurian and continued through the Devonian (Van der Pluijm 1993). Some believe the important phase of the Acadian orogeny, with regional metamorphism, intrusion and structural complexity is in the Givetian roughly, maybe beginning in the later Eifelian (Boucot et al. 2012).

Climates were relatively warm and dry as evidenced from reefal and evaporite deposits with an increase in sea level during the early Silurian time from prior deglaciation events. During the middle Silurian time, Michigan basin and a southern basin (north central Ohio) were covered by muddy carbonates bounded by high barrier reefs with siliciclastic muds accumulating further east. These elevated reefs prevented the flow of water in the basins thus resulting in evaporite deposits. Late Silurian interval was a time with highly evaporitic basins and thus reefs grew only in the southwest, Illinois and Indiana. Global climate was relatively warm and dry during the Devonian with an abrupt drop in temperature during the late Devonian, and sea level remained relatively high during this time. Shallow water carbonates in the Lower Devonian gave way to deeper water deposits in the Middle Devonian in Eastern America. There were no glaciers until the Late Devonian, when sea level eventually began to drop. Thus, the Acadian orogeny resulted in demolishing shaly deposits in eastern America and as a result of erosion, the sediments were continually being pushed to the outward edge forming the clastic Catskill wedge as seen in New York sections (Boucot 1981; Stanley 1999). These climatic events led to the faunal abundance and diversity during the Silurian and Devonian and tectonic events led to their mixing and migration across barriers. Thus, it is important to determine how morphological patterns of then existing well preserved brachiopods within this wide range of time responded to these environmental changes.

1.5 Ecological Interactions

Often, brachiopods are the host of preference for studying encrustation in the Paleozoic fossil record, usually because of the abundance of specimens and, in the case of many siliciclastic units, ease of removal from the matrix. Indeed, because of the abundance of data collected about brachiopods and their encrusters, episkeletobiont ecological relationships are fairly well understood for the Paleozoic record. Brachiopod hosts are useful for investigating influences on episkeletobiont preferences, such as substrate texture (Schneider and Webb 2004; Schneider and Leighton 2007; Rodland et al. 2004), size of host (Ager 1961; Kesling et al. 1980), antifouling strategies (Schneider and Leighton 2007), and host-switching during taxon loss across mass extinctions (Schneider and Webb 2004). Although other taxa have been utilized for similar investigations, brachiopods remain one of, if not the most, well-understood hosts for Paleozoic episkeletobionts. Less well understood is the morphological response of these Paleozoic brachiopod morphological shape patterns in response to encrustation rates within these ecological evolutionary subunits. Atrypide hosts preserving numerous episkeletobionts during this time provide a good case study for determining their life modes, filtering mechanisms and in studying their morphological shape pattern in response to the encrustation frequency.

The diversity of Paleozoic encrusting organisms peaked during the Devonian. Several clades of bryozoans and echinoderms were already established as organisms which attach to brachiopods prior to the Devonian; other encrusters, such as spirorbid worms, auloporid corals, and hederellids first appear or become abundant during this time. Cystoporate bryozoans, many of which are encrusters, achieve their maximum diversity during the Devonian, and trepostome bryozoans, which also include encrusting forms, also experience a resurgence at this time (McKinney and Jackson 1989). Moreover, encrustation frequency and host area covered by encrusters also increased during the Devonian (Taylor and Wilson 2003). The Devonian relationship between encruster and host represents a fundamental shift in Paleozoic encruster behavior. Prior to the Devonian, encrustation was most common on abiotic surfaces, such as hardgrounds and rocks (Taylor and Wilson 2003). These abiotic surfaces were common in Early Paleozoic sedimentological regimes, but with the decline in these hard substrates, epizoans were more limited in choice of substrate. Hence, most encruster occurrences in Devonian ecosystems were on the skeletal material of live and dead hosts (Taylor and Wilson 2003). The proposed research will capture the P3 EEU of the Silurian–Devonian time interval and investigate the evolutionary response of atrypide valve morphology in response to ecological interactions. Location of these episkeletobionts and their response to microornamentation within the 2 subfamilies were not determined precisely, thus life mode of the atrypide hosts and the encruster preference for subfamilies cannot be clearly stated herein.

1.6 Research Hypotheses

Using these data, 4 hypotheses were tested: (1) If these 2 subfamilies are closely related to each other with their internal morphologies closely similar, then we would expect insignificant morphological differences between samples from Early Silurian-Early Devonian (Atrypinae dominated) and Middle Devonian–Late Devonian (Variatrypinae dominated) time intervals; (2) If subfamily Variatrypinae was derived from the Atrypinae subfamily, and atrypide morphological shape varied with respect to their subfamilies, then we would expect samples from Early Silurian-Early Devonian (Atrypinae dominated data) and Middle Devonian-Late Devonian (Variatrypinae dominated data) form two separate clusters in a dendogram with a close link between the two clusters due to their observed similarity in internal and external morphological characters. Alternatively, if the morphological shape of these atrypide individuals varied in the two subfamilies, and they evolved in a gradual, directional manner, then we would expect samples close together in time to be more similar to one another than those more separated in time. (3) If brachiopod individuals of these 2 subfamilies are cosmopolitan, then we would expect significant difference in atrypide morphological shape within these subfamilies analyzed from different geographic localities. (4) If the encrustation frequency changes in an organism's life history over time, it is expected for them to change their morphological shape through time. If this is true, we expect a strong correlation between encrustation frequency/encrustation rate and morphological shape change through time.

References

Ager DV (1961) The epifauna of a Devonian spiriferid. Q J Geol Soc Lond 117:1–10

Bose R (2012a) A new morphometric model in distinguishing two closely related extinct brachiopod species. Hist Biol 24:1–10

Bose R (2012b) Quantitative analysis strengthens qualitative assessment: a case study of Devonian brachiopod species. Paläontologische Zeitschrift, Scientific Contributions to Palaeontology 86:1–8

Bose R (2012c) Biodiversity and evolutionary ecology of extinct organisms. Springer Verlag book series. p 214. ISBN 978-3-642-31720-0

Bose R, De A (2013) Quantitative evaluation reveals taxonomic over-splitting in extinct marine invertebrates: implications in conserving biodiversity. In: Proceedings of the National Academy of Sciences, India Section B: Biological Sciences 83:1–5

Boucot AJ (1983) Does evolution take place in an ecological vacuum? J Paleontol 57:1–30

Boucot AJ (1986) Ecostratigraphic criteria for evaluating the magnitude, character and duration of bioevents. In: Walliser OH (ed) Global bio-events. Lecture notes earth science, vol 8. Springer-Verlag, Berlin, pp 25–45

Boucot AJ (1990) Community evolution: its evolutionary and biostratigraphic cance. In: Miller W (ed) Paleocommunity temporal dynamics: the long-term development of multispecies assemblages, vol 5. Paleontological Society Special Publication, The Paleontological Society, Boulder, CO, pp 48–70

Boucot AJ (1981) Does Evolution take place in an ecological vacuum? J Palaeontol 57(1):1–30

Boucot AJ, Blodgett RB, Rohr DM (2012) Brachiopoda (Atrypidae), from upper Silurian strata of the Alexander terrane, northeast Chichagof Island, Alaska. Bull Geosci 87:261–267

Bowen ZP (1966) Intraspecific variation in the brachial cardinalia of *Atrypa reticularis*. J Paleontol 40:1017–1022

Brett CE, Baird GC (1995) Coordinated stasis and evolutionary ecology of Silurian to Middle Devonian faunas in the Appalachian Basin. In: Erwin DH, Anstey RL (eds) New approaches to speciation in the fossil record. Columbia University Press, New York, pp 285–315

Brett CE, Miller KB, Baird GC (1990) A temporal hierarchy of paleoecological processes within a Middle Devonian epeiric sea. In: Miller W (ed) Paleocommunity temporal dynamics: the long-term development of multispecies assemblages, vol 5. Paleontological Society Special Publication, The Paleontological Society, Boulder, CO, pp 178–209

Campbell KSW, Chatterton BDE (1979) Coelospira: do its double spires imply a double lophophore? Alcheringa 3:209–223

Carlson SJ, Leighton LR (2001) The Phylogeny and classification of rhynchonelliformea. In: Carlson SJ, Sandy MR (eds) Brachiopods ancient and modern. The Paleontological Society Special Publications, New Haven, pp 27–51

Copper P (1977) The Late Silurian brachiopod genus Atrypoidea. Geol Foren Stockholm Forh 99:10–26

Copper P (1967) Pedicle morphology in Devonian atrypid brachiopods. J Paleontol 41:1166–1175

Copper P (1973) New Siluro-Devonian atrypoid brachiopods. J Paleontol 47:484–500

Copper P (1996) New and revised genera of Wenlock-Ludlow Atrypids (Silurian Brachiopoda) from Gotland, Sweden, and the United Kingdom. J Paleontol 70:913–923

Copper P (1997) Reefs and carbonate productivity: Cambrian through Devorían. Proceedings of the 8th International Coral Reef Symposium, Panama 2:1623–1630

Copper P (2001) Radiations and extinctions of atrypide brachiopods: Ordovician–Devonian. In Brunton CHC, Cocks LRM, Long SL (eds) Brachiopods past and present. Natural History Museum, London, pp 201–211

Copper P (2002) Atrypida. In: Kaesler RL (ed) Brachiopoda (revised), part H of treatise on invertebrate paleontology. The Geological Society of America, Inc. and the University of Kansas, Boulder, Colorado and Lawrence, Kansas; Publications, vol 4, pp 1377–1474

Copper P (2004) Silurian (Late Llandovery-Ludlow) Atrypid Brachiopods from Gotland, Sweden, and the Welsh Borderlands, the Great Britain. National Research Council of Canada Research Press, Ottawa

Day J (1998) Distribution of latest Givetian-Frasnian Atrypida (Brachiopoda) in central and western North America. Acta Palaeontol Pol 43:205–240

Day J, Copper P (1998) Revision of latest Givetian-Frasnian Atrypida (Brachiopoda) from central North America. Acta Palaeontol Pol 43:155–204

Eldredge N, Gould SJ (1972) Punctuated equilibria: an alternative to phyletic gradualism. In: Schopf TJM (ed) Models in paleobiology. Freeman, San Francisco, pp 82–115

Fenton CL, Fenton MA (1930) Studies on the Genus *Atrypa*. Am Midl Nat 12:1–18

Gingerich PD (1976) Paleontology and phylogeny: patterns of evolution at the species level in early tertiary mammals. Am J Sci 276:1–28

Goldman D, Mitchell CE (1990) Morphology, systematics, and evolution of Middle Devonian Ambocoeliidae (Brachiopoda), western New York. J Paleontol 64:79–99

Gould SJ, Eldredge N (1977) Punctuated Equilibria: the tempo and mode of evolution reconsidered. Paleobiology 3:115–151

Gradstein F, Ogg J, Smith A (2005) A geologic time scale 2004. Geol Mag 142:633–634

Hancock A (1858) On the organization of the Brachiopoda. Philos Trans R Soc Lond 148:791–870

Haney RA, Mitchell CE (1998) Multivariate analysis of geographic variation in valve shape of the brachiopod *Sowerbyella curdsvillensis*. Abstracts with programs. Geol Soc Am 30:35

Haney RA, Mitchell CE, Kim K (2001) Geometric morphometric analysis of patterns of shape change in the Ordovician brachiopod *Sowerbyella*. Palaios 16:115–125

Holterhoff PE (1996) Crinoid biofacies in Upper Carboniferous cyclothems, midcontinent North America: faunal tracking and the role of regional processes in biofacies recurrence. Palaeogeogr Palaeoclimatol Palaeoecol 127:47–81

Hunt G (2007) The relative importance of directional change, random walks, and stasis in the evolution of fossil lineages. Proc Natl Acad Sci 104:18404–18408

Isaacson PE, Perry DG (1977) Biogeography and morphological conservatism of *Tropidoleptus* (Brachiopoda, Orthida) during the Devonian. J Paleontol 51:1108–1122

Ivany LC, Brett CE, Wall HLB, Wall PD, Handley JC (2009) Relative taxonomic and ecologic stability in Devonian marine faunas of New York State: a test of coordinated stasis. Paleobiology 35:499–524

Jones B (1974) A biometrical analysis of *Atrypella foxi* n. sp. from the Canadian Arctic. J Paleontol 48:963–977

Jones B (1977) Variation in the Upper Silurian brachiopod *Atrypella phoca* (Salter) from the Somerset and Prince of Wales Islands. J Paleontol 51:459–479

Jones B (1982) Paleobiology of the Upper Silurian Brachiopod Atrypoidea. J Paleontol 56:912–923

Kesling RV, Hoare RD, Sparks DK (1980) Epizoans of the Middle Devonian brachiopod *Paraspirifer bownockeri*: their relationships to one another and to their host. J Paleontol 54:1141–1154

Lieberman BS, Brett CE, Elredge N (1995) A study of stasis in two species lineages from the Middle Devonian of New York State. Paleobiology 21:15–27

Mckinney FK, Jackson JBC (1989) Bryozoan evolution. University of Chicago Press, Chicago

Morris PJ (1995) Coordinated stasis and ecological locking. Palaios 10:101–102

Morris PJ, Ivany LC, Schopf KM, Brett CE (1995) The challenge of paleoecological stasis: reassessing sources of evolutionary stability. Proc Natl Acad Sci 92:11269–11273

Rodland DL, Kowalewski M, Carroll M, Simões MG (2004) Colonization of a 'Lost World': encrustation patterns in modern subtropical Brachiopod assemblages. Palaios 19(4):381–395

Rudwick MJS (1958) Filter-feeding mechanisms in some brachiopods from New Zealand. Zool J Linnean Soc 44:592–615

Rudwick MJS (1960) The feeding mechanisms of spire-bearing fossil brachiopods. Geol Mag 97:369–383

Rudwick MJS (1965a) Ecology and paleoecology. In: Moore RC (ed) Treatise in invertebrate paleontology, Part H–Brachiopoda, vol 1, pp H 199–H214

Rudwick MJS (1970) Living and fossil brachiopods. Hutchinson University Library Press, London 199

Schneider CL, Webb A (2004) Where have all the encrusters gone? Encrusting organisms on Devonian versus Mississippian brachiopods. Geol Soc Am Abstr 36:111

Schneider CL, Leighton LR (2007) The influence of spiriferide micro-ornament on Devonian epizoans. Geol Soc Am Abstr 39:531

Sheehan PM (1991) Patterns of synecology during the Phanerozoic. In: Dudley EC (ed) The unity of evolutionary biology. Dioscorides Press, Portland, pp 103–118

Sheehan PM (1996) A new look at ecologic evolutionary units (EEUs). Palaeogeogr Palaeoclimatol Palaeoecol 127:21–32

Sheldon PR (1987) Parallel gradualistic evolution of Ordovician trilobites. Nature 330:561–563

Sheldon PR (1990) Shaking up evolutionary patterns. Nature 345:772

Sheldon PR (1996) Plus qa change—a model for stasis and evolution in different environments. Palaeogeogr Palaeoclimatol Palaeoecol 127(209):227

Stanley SM (1999) Earth system history. John Hopkins University. Freeman and Co., New York

Taylor PD, Wilson MA (2003) Palaeoecology and evolution of marine hard substrate communities. Earth Sci Rev 62:1–103

Van der Pluijm B (1993) Paleogeography, accretionary history, and tectonic scenario; a working hypothesis for the Ordovician and Silurian evolution of the Northern Appalachians. Special Paper—Geological Society of America, vol 275, pp 27–40

Webber AJ, Hunda BR (2007) Quantitatively comparing morphological trends to environment in the fossil record (Cincinnatian series; Upper Ordovician). Evolution 61:1455–1465

Chapter 2
Materials and Method

2.1 Geometric Morphometrics

Geometric morphometrics is the analysis of geometric landmark coordinates points on specific parts of an organism (Bookstein 1991). Morphometric analyses were based on the use of landmarks to capture shape; landmarks are points representing the same location on each specimen, and can be assigned to three general categories (Bookstein 1991): Type I—discrete juxtapositions (e.g., meeting of three structures), Type II—functional equivalents (e.g., tips of extrusions and maximum curvatures) and Type III—extremal points (e.g., tip of beak). In this study, type I and II landmarks were used, as they are most likely to capture biologically meaningful shape change (Fig. 2.1). When selecting landmarks for analyses, we selected points that characterized not only body shape accurately, but also represent some aspect of the inferred ecological niche. While landmarks were not biologically homologous, they represent discrete points that correspond among forms (sensu Bookstein 1991), which is appropriate for analyses attempting to capture shape changes or function (as opposed to describing phylogenetic relationships). For all individuals, measurements were taken on the entire specimen (brachial and pedicle valve. Once coordinates were obtained, geometric morphometrics were applied to landmarks to convert them into Bookstein shape coordinates (see Bookstein 1991 for a full description of Bookstein shape coordinate equations and methods), thus rotating, translating and scaling all landmarks, while maintaining their geometric relationships; these scaled coordinates were used in all analyses.

2.2 Data Set

All Silurian-Devonian atrypide taxa from ENA were identified based on external morphological characters and ornamentation. An atrypide individual was CT-scanned to determine the internal morphology of a taxa from Variatrypinae subfamily using a micro-scanner at 36 μm resolution (Fig. 2.2). We tested morphological variation using a total of 1,551 brachial valves and 1554 pedicle valves of well preserved atrypide brachiopods from the Middle Paleozoic Era of Eastern North American outcrops measured from six successive time units from

R. Bose, *Palaeobiology of Middle Paleozoic Marine Brachiopods*,
SpringerBriefs in Earth Sciences, DOI: 10.1007/978-3-319-00194-4_2,
© The Author(s) 2013

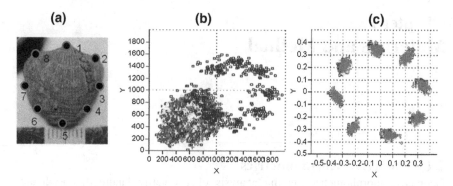

Fig. 2.1 Size, translation and rotation removed by rescaling through Procrustes superimposition of landmarks. **a** Eight landmarks on pedicle valve **b** Original shape data **c** Procrustes aligned data after superimposition

Fig. 2.2 Dorsal valve of *Desquamatia sp.* (Variatrypinae subfamily) from Middle Devonian Hamilton Silver Creek, Indiana showing the D-shaped brachidium and jugal processes

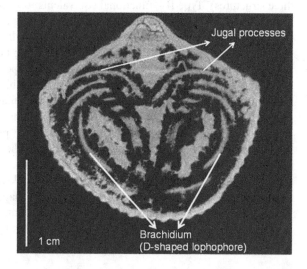

7 EE subunits, 1,193 of which were used to assess geographic variation within 3 time intervals (Middle Silurian = 329, Early Devonian = 406 and Middle Devonian = 458). Stratigraphic information for samples collected from each time unit is described in Table 2.1. Samples chosen for this study were restricted to the ENA region and they belong to a varied environmental setting from shales to carbonates to mixed carbonate-siliciclastic environment (Table 2.1). Geographic location chosen and respective sample count is detailed in a world map (Table 2.1; Fig. 2.3).

These specimens are housed in the Invertebrate Paleontology Collections of the American Museum of Natural History, Yale Peabod y Museum, New York State Museum and Indiana University Paleontology Collections.

Table 2.1 Specimen count and their geographic location, depositional environment and EE 1185 subunits from six time intervals—Early Silurian, Middle Silurian, Late Silurian, Early Devonian, 1186 Middle Devonian and Late Devonian

Time units	Stages	EE subunits	Specimen count (1551-B, 1554-P)	Formation and localities	Environment
DEVONIAN					
Late Devonian	Frasnian	11 subunit	Iowa Basin Total = 123	Cerro Gordo member, Lime Creek shale, Hackberry grove, North-central IA	Mid-outer carbonate shelf: restricted and open marine oxic facies
Middle Devonian	Uppermost Eifelian–Middle Late Givetian Lower-Mid Eifelian	Hamilton (6–7 Myr)	146—Michigan Basin (48–43 from NYSM and 5 from AMNH) Appalachian Basin (146 new—Hamilton Grp) 63—Cincinnatian Arch 33—Hamiltonian, Fulton, Missouri.	146—Traverse,NE MI; 194?—Hamilton, NY; 63—Hamilton Silver Creek, IN; 33—Hamiltonian, MO	MI—mixed setting New York—siliciclastics bordered by carbonate shelf IN—carbonate MO—carbonate
		Onondaga (5–6 Myr)	9—Dundee Limestone (Ohio) 35—Onondaga Limestone (New York) Total = **458**	9—Ohio 35—Onondaga, NY	OH—siliciclastic NY—carbonate
Early Devonian	Emsian Loch kovian	Schoharie (5 Myr)	48—Appalachian Basin	152—Linden Grp, Helderbergian, TN; 34—Schoharie Grit, New Scotland Limestone, Western New York 49—Keyser Lst, WV 31—Keyser Lst, MD 31—Lr Helderberg Group, NY 74—Yellow shale below limestone, Oklahoma	TN-siliciclastic Western NY – shallow carbonates and siliciclastics. WV = carbonate MD = carbonate NY = carbonate Carbonate
		Helderberg (6 Myr)	32—Oklahoma Basin Total = 406	32—Hunton Lst, Oklahoma	Carbonate ramp ('Hunton Ramp') on the margin of the Oklahoma Aulacogen and Ouachita Trough

(continued)

Table 2.1 (continued)

Time units	Stages	EE subunits	Specimen count (1551-B, 1554-P)	Formation and localities	Environment
Upper Silurian	Ludlovian		96—Oklahoma Basin Total = 94(B), 96(P)	96—Henryhouse Limestone, Hunton group	Carbonate
Middle Silurian	Wenlock	Upper Clinton-Lockport (7-8 Myr)	152—Tennessee, App. Basin 69—Cincinnatian Arch between Illinois and Michigan Basin 107—Appalachian Basin Total = 329	152—Wayne Fm., Tennessee (different localities) 69—Waldron, N.Grp., Indiana 107—Lockport, New York	Dudley-Carbonate shelves: mid-platform environments between storm- and fair-weather wave-base Indiana—carbonates. Siliciclastics
Early Silurian	Llandovery	Lower Clinton (4 Myr)	141—Hudson Basin (Check specimen nos 1, 114) Total = 141(b), 142(p)	Jupiter Fm, Anticosti Island, Canada	Middle shelf mudstones and tempestites between storm- and fair-weather wave-base

SILURIAN

Fig. 2.3 Sampled localities for atrypide brachiopods in Eastern North America. Filled triangle in 1,155 black indicates Silurian localities. Early Silurian = 141 (Canada); Middle Silurian = 329 (Indiana = 69; 1,156 Tennessee = 152; New York = 108), Late Silurian = 96 (Oklahoma). Filled square in black indicates Devonian 1,157 localities. Early Devonian = 406 (Maryland = 31; New York = 65; Oklahoma = 106; Tennessee = 154; West 1158 Virginia = 50); Middle Devonian = 458 (Indiana = 63; Michigan = 210; Missouri = 33; New York = 144; 1159 Ohio = 9); Late Devonian = 123 (Iowa)

Data collected for these atrypide individuals were analysed using geometric morphometrics. Eight landmarks were used to represent the specific parts of the atrypide valves. Procrustes analysis was performed to remove size, translation and rotation effects from digital photographs. Shell shape patterns were determined using principal component analysis. Multivariate analysis was performed to analyse shape variation in atrypides over time and space. Alternatively, Euclidean cluster analysis was performed to determine the relatedness between the two subfamilies that persisted over different time intervals. Cluster analysis is also useful in determining the geographic links between Middle Silurian, Early Devonian and Middle Devonian atrypide individuals collected from different geographic localities. Morphological distances in atrypide samples between different time intervals and geographic locations were noted from similarity and distance indices. Mean shape morphological trend was constructed using the Principal Component scores of these measured shapes to test whether morphological change or stasis was common in the P3 EEU. Average Procrustes distance was also calculated for both brachial and pedicle valves to determine the comparative rate of morphological shape change over time and space. Thin Plate Spline visualisation plots were constructed to determine the variation in the mean shape of the valve morphology in the Atrypinae (Early Silurian–Early Devonian) and Variatrypinae (Middle Devonian–Late Devonian) subfamilies.

Encrustation frequency was determined by counting the number of valves encrusted from each time unit for both brachial and pedicle valves. Statistical correlation between encrustation rate and morphological change (similarity and distance indices) was determined temporally using this formula:

$$E_A \text{ (Average encrustation rate)} = \sum E_e / \sum H_e / N \tag{1}$$

where E_e is the total number of encrusters, H_e is the number of encrusted host valves and N is the total number of specimens in each time unit.

Reference

Bookstein FL (1991) Morphometric tools for landmark data: geometry and biology. Cambridge U. Press, Cambridge

Chapter 3
Results

3.1 Temporal Variation

Principal component analysis of atrypide individuals from the 2 subfamilies of Eastern North America region shows some morphological variation and some overlap among the six clustered groups of atrypides based on six time units. Multivariate analysis performed on morphological shape of atrypides from six time intervals is illustrated in Fig. 3.1.

MANOVA and bootstrap test suggests significant shape differences between different time horizons ($p \leq 0.01$), suggesting at least some variation in atrypide morphology in the two subfamilies (Table 3.1) thus partially refuting our hypothesis 1 of morphological relatedness of the Atrypinae subfamily to the Variatrypinae subfamily.

Euclidean based cluster analysis for brachial valve samples shows Variatrypinae dominated samples (Middle–Late Devonian) forms one separate cluster with some Atrypinae dominated samples (Middle Silurian and Early Devonian) while the other cluster is formed by rest of the Atrypinae dominated samples (Early and Late Silurian). Euclidean based cluster analysis shows similar results for pedicle valve samples with one cluster dominated by Variatrypinae dominated samples (Middle–Late Devonian) and some Atrypinae dominated samples (Early–Middle Silurian) and the other cluster exclusively dominated by rest of the Atrypinae samples (Late Silurian–Early Devonian). This refutes our hypothesis partially as the two subfamilies do not form a separate cluster as expected. In addition, these dendograms depict samples widely separated in time are more similar than those more close to each other (Fig. 3.2). On average, similarity and distance indices for morphological distances measured between time units suggest relatively greater morphological distances between successive time units as compared to lesser morphological distances between time units widely separated from each other (Table 3.2).

3.2 Spatial Variation

Multivariate analysis indicates significant geographic differentiation in Middle Silurian and Early Devonian samples (Tables 3.3 and 3.4; $p \leq 0.01$) while minor geographic differentiation in Middle Devonian samples (Table 3.5; $p \leq 0.01$).

R. Bose, *Palaeobiology of Middle Paleozoic Marine Brachiopods*,
SpringerBriefs in Earth Sciences, DOI: 10.1007/978-3-319-00194-4_3,
© The Author(s) 2013

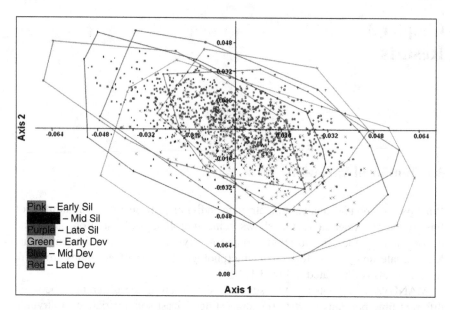

Fig. 3.1 Canonicate variate analysis for atrypides from 6 different time intervals from Eastern North American realm [Early Sil (*solid square* in *pink*), Mid Sil (*vertical bar* in *maroon*), Late Sil (*unfilled diamond* in *purple*), Early Dev (*cross* in *green*), Mid Dev (*unfilled square* in *blue*), Late Dev (*Plus* in *red*)]

Table 3.1 'p' values showing significant variation in valve morphological shape in atrypides through time

Time units	'p' values (brachial valve)	'p' values (pedicle valve)
Late Dev–Middle Dev	0.000	0.000
Late Dev–Early Dev	0.000	0.000
Late Dev–Late Sil	0.000	0.000
Late Dev–Middle Sil	0.000	0.000
Late Dev–Early Sil	0.000	0.000
Middle Dev–Early Dev	0.000	0.000
Middle Dev–Late Sil	0.000	0.000
Middle Dev–Middle Sil	0.000	0.000
Middle Dev–Early Sil	0.000	0.000
Early Dev–Late Sil	0.000	0.000
Early Dev–Middle Sil	0.000	0.000
Early Dev–Early Sil	0.000	0.000
Late Sil–Middle Sil	0.000	0.000
Late Sil–Early Sil	0.000	0.000
Middle Sil-Early Sil	0.000	0.000

Cluster analysis reveals Middle Silurian samples from Tennessee and Indiana are more closely related to each other than New York for both brachial and pedicle valves (Fig. 3.3). Same analysis reveals Early Devonian samples from

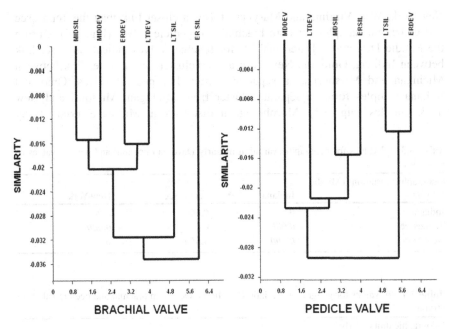

Fig. 3.2 Morphological links in time intervals for both brachial and pedicle valves for atrypides

Table 3.2 Morphological distance (similarity and distance indices) between six time intervals of 1190 Eastern North American atrypide morphology for (a) brachial and (b) pedicle valves

Time units	Morphological distance (brachial valve)	Morphological distance (pedicle valve)
Late Dev–Middle Dev	0.02	0.03
Late Dev–Early Dev	0.02	0.03
Late Dev–Late Sil	0.03	0.03
Late Dev–Middle Sil	0.03	0.02
Late Dev–Early Sil	0.04 (average for widely separated time units)	**0.02 (lesser for widely separated time units)**
Middle Dev–Early Dev	0.02	0.03
Middle Dev–Late Sil	0.03	0.04
Middle Dev–Middle Sil	0.02	0.02
Middle Dev–Early Sil	0.03	0.02
Early Dev–Late Sil	0.03	0.01
Early Dev–Middle Sil	0.02	0.03
Early Dev–Early Sil	0.04	0.02
Late Sil–Middle Sil	**0.04 (greater for successive time units)**	0.03
Late Sil–Early Sil	**0.05 (greater for successive time units)**	0.03
Middle Sil–Early Sil	0.03	0.02

New York, West Virginia and Maryland forms a closer link than the Tennessee and Oklahoma samples for both brachial and pedicle valves (Fig. 3.4). During the Middle Devonian, cluster analysis for brachial valves indicate a closer link between Indiana, Ohio and New York as one cluster and a closer link between Michigan and Missouri as a separate cluster. For pedicle valves, Ohio and Indiana samples forms a separate cluster from Michigan, Missouri and New York samples (Fig. 3.5). Morphological distances in Middle Silurian, Early

Table 3.3 'p' values for geographic variation in Early Devonian brachial and pedicle valves of atrypides

Geographic variations (Middle Silurian)	Indiana	Tennessee	New York
Indiana	**0**	*0.000*	*0.000*
Tennessee	*0.000*	**0**	*0.000*
New York	*0.000*	*0.000*	**0**

Table 3.4 'p' values for geographic variation in Middle Devonian brachial and pedicle valves of atrypides

Geographic units (Early Devonian)	'p' (brachial valves)	'p' (pedicle valves)
Maryland–New York	0.022	*0.001*
Maryland–West Virginia	*0.001*	0.087
Maryland–Tennessee	*0.000*	*0.002*
Maryland–Oklahoma	*0.007*	*0.000*
New York–West Virginia	*0.000*	*0.000*
New York–Tennessee	*0.000*	*0.000*
New York–Oklahoma	*0.010*	*0.000*
West Virginia–Tennessee	*0.000*	*0.003*
West Virginia–Oklahoma	*0.000*	*0.000*
Tennessee–Oklahoma	*0.000*	*0.000*

Table 3.5 'p' values for geographic variation in Middle Silurian brachial and pedicle valves of atrypides

Geographic units (Middle Devonian)	'p' (brachial valves)	'p' (pedicle valves)
Michigan–Indiana	*0.000*	*0.000*
Michigan–New York	*0.000*	*0.000*
Michigan–Ohio	1.000	0.100
Michigan–Missouri	*0.000*	*0.000*
Indiana–New York	*0.000*	*0.000*
Indiana–Ohio	0.558	0.407
Indiana–Missouri	*0.000*	*0.000*
New York–Ohio	1.000	1.000
New York–Missouri	*0.001*	*0.002*
Ohio–Missouri	0.016	0.096

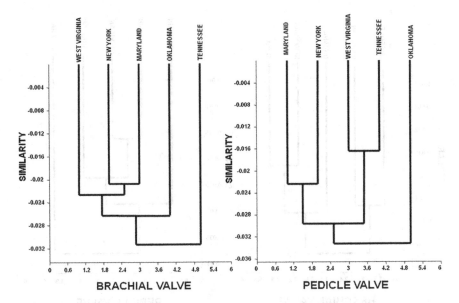

Fig. 3.3 Morphological links in biogeographic locations in Eastern North America for Middle Silurian brachial and pedicle valves for atrypides

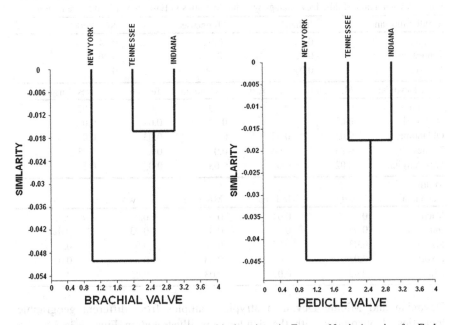

Fig. 3.4 Morphological links in biogeographic locations in Eastern North America for Early Devonian brachial and pedicle valves for atrypides

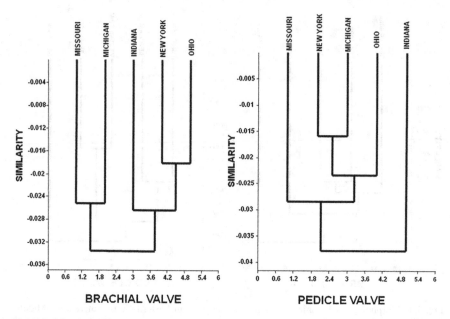

Fig. 3.5 Morphological links in biogeographic locations in Eastern North America for Middle Devonian brachial and pedicle valves for atrypides

Table 3.6 Morphological distance in brachial valves of atrypide samples from Middle Silurian, Early Devonian and Middle Devonian geographic localities in Eastern North America region

Middle Silurian	Indiana		Tennessee		New York
Indiana	0		0.02		0.04
Tennessee	0.02		0		0.06
New York	0.04		0.06		0

Early Devonian	Maryland	New York	Oklahoma	Tennessee	West Virginia
Maryland	0	0.02	0.03	0.03	0.02
New York	0.02	0	0.02	0.04	0.02
Oklahoma	0.03	0.02	0	0.03	0.03
Tennessee	0.03	0.04	0.03	0	0.03
West Virginia	0.02	0.02	0.03	0.03	0

Middle Devonian	Michigan	Indiana	Missouri	New York	Ohio
Mich	0	0.04	0.03	0.03	0.02
Ind	0.04	0	0.05	0.03	0.02
Misso	0.03	0.05	0	0.03	0.03
NYork	0.03	0.03	0.03	0	0.02
Ohio	0.02	0.02	0.03	0.02	0

Devonian and Middle Devonian atrypide samples from different geographic localities also comply with the dendograms illustrated in Figs. 3.3, 3.4, 3.5 (Tables 3.6 and 3.7).

Table 3.7 Morphological distance in pedicle valves of atrypide samples from Middle Silurian, Early Devonian and Middle Devonian geographic localities in Eastern North America region

Middle Silurian	Indiana	Tennessee	New York		
Indiana	0	0.02	0.04		
Tennessee	0.02	0	0.05		
New York	0.04	0.05	0		

Early Devonian	Maryland	New York	Oklahoma	Tennessee	West Virginia
Maryland	0	0.02	0.04	0.03	0.02
New York	0.02	0	0.03	0.03	0.03
Oklahoma	0.04	0.03	0	0.03	0.04
Tennessee	0.03	0.03	0.03	0	0.02
West Virginia	0.02	0.03	0.04	0.02	0

Middle Devonian	Michigan	Indiana	Missouri	New York	Ohio
Mich	0	0.03	0.03	0.02	0.02
Ind	0.03	0	0.04	0.03	0.05
Misso	0.03	0.04	0	0.02	0.03
NYork	0.02	0.03	0.02	0	0.02
Ohio	0.02	0.05	0.03	0.02	0

3.3 Mean Morphological Shape

Morphological mean shape trend constructed from Principal component scores for brachial valves indicate some morphological shape variation between Early and Late Silurian samples within the Atrypinae subfamily while those from Middle Silurian of the Atrypinae subfamily overlap with the Early–Late Devonian samples dominated by both Atrypinae (Early Devonian) and Variatrypinae (Middle–Late Devonian) samples (Fig. 3.6).

For pedicle valves, the mean shape trend remains the same for Early Silurian–Late Silurian samples with clear deviation in Early Devonian samples from the lower Silurian samples suggesting almost no morphological variation within Silurian Atrypinae subfamily. Morphological overlap is significant between Early and Late Devonian samples for pedicle valves with some deviation in trend in Middle Devonian samples further suggesting no variation between Atrypinae and Variatrypinae subfamilies but some variation within Variatrypinae subfamily itself due to fluctuating trend observed between Middle and Late Devonian samples that is dominated by individuals from Variatrypinae subfamily. Overall morphological overlap between lowermost and uppermost occurrences is prominent for both brachial and pedicle valves for the atrypide group (Fig. 3.6).

A general observation shows an overall significant size increase from smaller shells in the Silurian to larger ones in the Devonian, with the Lochkovian being "intermediate." This size difference applies to many of the Silurian-Devonian families including the spiriferids, chonetids, atrypoids, atrypaceans, and some pentameroids (mostly among the gyidulids) (Boucot et al. 2012).

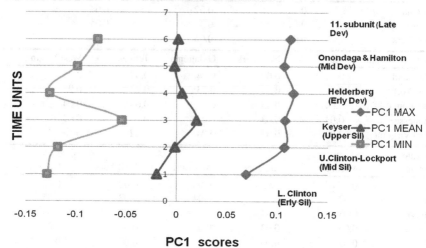

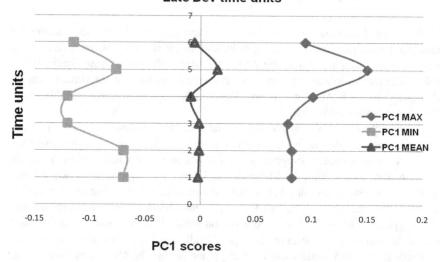

Fig. 3.6 Mean PCA trend with a minimum and maximum trend for atrypide morphological shape for 1172 both brachial and pedicle valves for the six time units (*1* Early Silurian, *2* Middle Silurian, *3* Late 1173 Silurian, *4* Early Devonian, *5* Middle Devonian, *6* Late Devonian)

Average Procrustes distance over time suggest morphological shape change in brachial and pedicle valves is closely similar (Table 3.7). Average Procrustes distance in space is closely similar for both brachial and pedicle valves which further

Table 3.8 Average procrustes distance for brachial and pedicle valves in time

	Brachial valve	Pedicle valve
Time	0.025	0.028

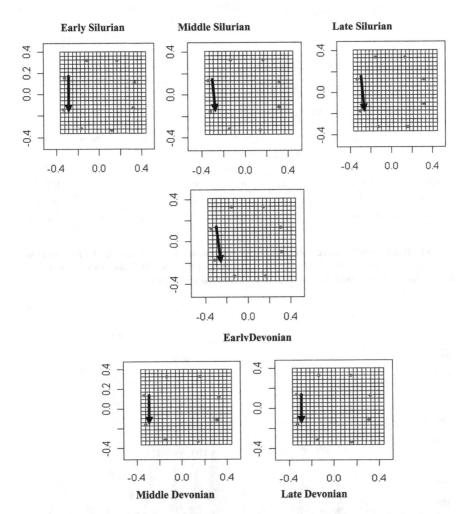

Fig. 3.7 Mean shape trend in specific parts of brachiopod valve morphology: TPS visualisation plots for brachial valve mean shape. **a** Atrypinae subfamily (Early Silurian, Middle Silurian, Late Silurian, Early Devonian). **b** Variatrypinae subfamily (Middle and Late Devonian)

suggest morphological shape change did not differ much with respect to valves (Table 3.8). Thin Plate Spline visualisation plots indicate slight deflection in posterior left of the brachial valve within the Atrypinae subfamily dominating the Early Silurian–Early Devonian time period while morphological shape of the brachial valve remains relatively stable within the Variatrypinae subfamily (Fig. 3.7).

Encrustation frequency in Brachial and Pedicle valves

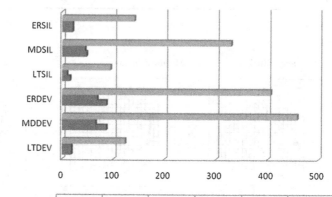

	LTDEV	MDDEV	ERDEV	LTSIL	MDSIL	ERSIL
▦ Valve count (brachial and pedicle)	123	458	406	94	329	141
▦ Encrusted pedicle valve	18	65	68	9	44	20
▦ Encrusted brachial valve	18	86	86	15	47	20

Fig. 3.8 Brachial and pedicle valve encrustation frequency from six time units. Encrustation frequency here refers to the total number of valves encrusted in each time unit. Valve count refers to the total number of specimens studied for both brachial and pedicle valves

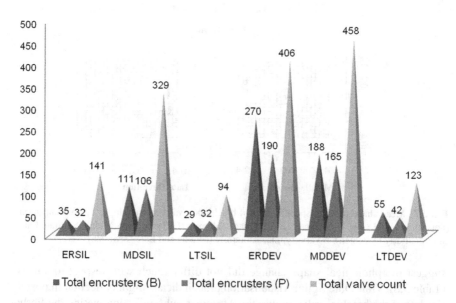

Fig. 3.9 Encruster count for both brachial and pedicle valves from each time unit. Numbers refer to the total count of brachial valve encrusters, pedicle valve encrusters and total valves examined

Table 3.8 Average procrustes distance for brachial and pedicle valves in time

	Brachial valve	Pedicle valve
Time	0.025	0.028

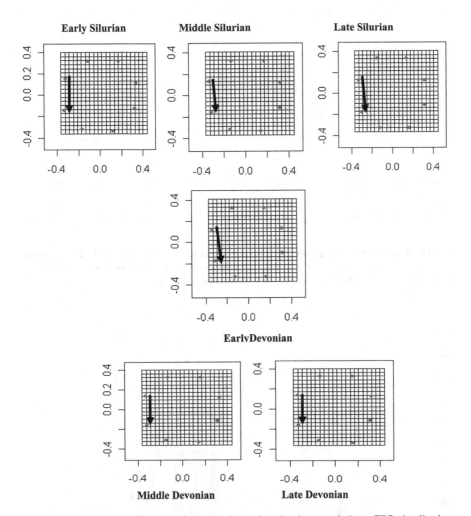

Fig. 3.7 Mean shape trend in specific parts of brachiopod valve morphology: TPS visualisation plots for brachial valve mean shape. **a** Atrypinae subfamily (Early Silurian, Middle Silurian, Late Silurian, Early Devonian). **b** Variatrypinae subfamily (Middle and Late Devonian)

suggest morphological shape change did not differ much with respect to valves (Table 3.8). Thin Plate Spline visualisation plots indicate slight deflection in posterior left of the brachial valve within the Atrypinae subfamily dominating the Early Silurian–Early Devonian time period while morphological shape of the brachial valve remains relatively stable within the Variatrypinae subfamily (Fig. 3.7).

Encrustation frequency in Brachial and Pedicle valves

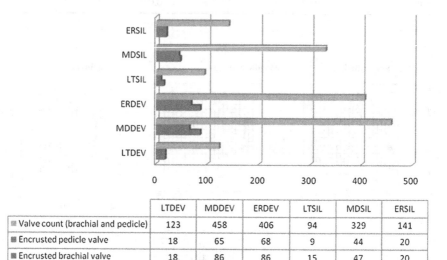

	LTDEV	MDDEV	ERDEV	LTSIL	MDSIL	ERSIL
▨ Valve count (brachial and pedicle)	123	458	406	94	329	141
▪ Encrusted pedicle valve	18	65	68	9	44	20
▪ Encrusted brachial valve	18	86	86	15	47	20

Fig. 3.8 Brachial and pedicle valve encrustation frequency from six time units. Encrustation frequency here refers to the total number of valves encrusted in each time unit. Valve count refers to the total number of specimens studied for both brachial and pedicle valves

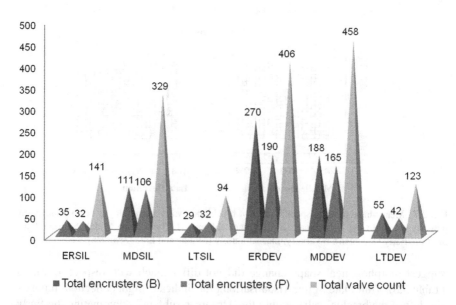

Fig. 3.9 Encruster count for both brachial and pedicle valves from each time unit. Numbers refer to the total count of brachial valve encrusters, pedicle valve encrusters and total valves examined

Table 3.9 Encrustation rate {(Σencrustation count/Σencrusted valves)/Total valves} effects on morphological change in brachial and pedicle valves in atrypide subfamilies of P3 EEU

	Encrustation rate	Morphological distance
Brachial		
Late Dev	0.0248	0.02
Mid Dev	0.0048	0.02
Early Dev	0.0078	0.03
Late Sil	0.0206	0.04
Middle Sil	0.0071	0.03
R	0.133039201	
R^2	0.017699429	Correlation = **2 %**
Pedicle		
Late Dev	0.019	0.03
Mid Dev	0.0055	0.03
Early Dev	0.0069	0.01
Late Sil	0.037	0.03
Middle Sil	0.0073	0.02
R	0.508416655	
R^2	0.258487495	Correlation = **26 %**

Table 3.10 Average procrustes distance for brachial and pedicle valves in three space units

	Brachial valve	Pedicle valve
Middle Devonian	0.040	0.037
Early Devonian	0.027	0.029
Middle Silurian	0.030	0.029

Table 3.11 'p' value indicates insignificant difference of encrustation rates in brachial and pedicle valves through time

Comparative encrustation rates in brachial and pedicle valves through time		
p value	Brachial valve	Pedicle valve
Late Dev	0.0248	0.019
Mid Dev	0.0048	0.0055
Early Dev	0.0078	0.0069
Late Sil	0.0206	0.037
Middle Sil	0.0071	0.0073
Early Sil	0.0125	0.0113
	p value	**0.39**

3.4 Encrustation Versus Morphological Shape

Brachial valves are encrusted more often as compared to pedicle valves in the six time units (Fig. 3.8). Total encrusters (encrustation frequency) are also higher in brachial valves in each time unit with the exception of Late Silurian interval (Fig. 3.9). However, encrustation rate is insignificant between the two valves

($p \leq 0.01$) (Table 3.9). Statistical correlation is weak ($R^2 = 0.01$) between brachial valve morphological distance and encrustation rate in each time unit while slightly strong between pedicle valve morphological distance and encrustation rate ($R^2 = 0.26$) (Tables 3.10 and 3.11).

Reference

Boucot AJ, Blodgett RB, Rohr DM (2012) Brachiopoda (Atrypidae), from upper Silurian strata of the Alexander terrane, northeast Chichagof Island, Alaska. Bull Geosci 87:261–267

Chapter 4
Discussion

4.1 Morphology

Morphological shape change patterns show relatively greater variation within members of the Atrypinae subfamily than those from the Variatrypinae subfamily for brachial valves while relatively less variation within Atrypinae subfamily than the Variatrypinae subfamily for pedicle valves suggesting minor distinction in morphological variation between the two subfamilies. Copper (1973) outlined a few new Siluro-Devonian species-groups within the Atrypidae order from eastern North American forms from Anticosti island and the interior Devonian carbonate platform. Atrypides possess ribs normally interrupted by growth lamellae, with a generally small pedicle foramen, small interarea, lacking deltidial plates occasionally but usually with small, disjunct hollow deltidial plates (rarely associated with collars), with disjunct jugal processes tipped by small plates, generally poorly developed dental cavities. The shell surface may be projected as frills or spines with their crowding along the anterior margins or may even evolve in directions towards loss of growth lamellae and development of continuous ribs.

4.2 Climate Setting in the Silurian and Devonian

Berry and Boucot (1970) have demonstrated clearly that the Silurian was a period in the Paleozoic during which shallow marine carbonate deposition was extremely widespread. Climates also must have been very warm during Wenlock time in parts of Michigan, Ontario and Ohio and Indiana because of the presence of abundant patch reefs (Lowenstam 1957). However, in the Llandoverian reefs were less widespread and the richly fossiliferous strata of that age from Anticosti island, Quebec, are poor in coral faunas. Thus there was probably a gradual warming up during the Llandoverian, with the onset of reef growth later on in the Wenlockian (Copper 1973, 2004). By Ludlow-Pridoli time the onset of evaporitic conditions was observed in eastern North America, with some of the sea lanes closing or disappearing (the Caledonian sea lane), and a barrier of some sort developing between European and North American areas, simply due

R. Bose, *Palaeobiology of Middle Paleozoic Marine Brachiopods*,
SpringerBriefs in Earth Sciences, DOI: 10.1007/978-3-319-00194-4_4,
© The Author(s) 2013

to uplift. This phenomenon seems to have continued during much of the Lower Devonian which saw the eventual extinction of the pelagic graptolite communities (Copper 1973). Eastern North Americas also continued with a carbonate platform setting. The faunal separation between the Appalachian and Hercynian sealanes persisted through the Middle Devonian when carbonates were widespread in both regions (Copper 1973, Fig. 2). Sometime during the late Givetian, or early Frasnian (Upper Devonian) full marine connections were reestablished between the Old World provinces and the Appalachians (Copper 1973, Fig. 3). At the end of Frasnian time, most shallow marine areas of the eastern North American platform and western Europe began to rise, producing black shales or disconformities locally. Eventually, the muddy bottom dwelling and stenohaline atrypides could not adapt to the evaporitic, euxinic and coarser clastic sedimentary bottoms widespread in the region, and died out (Copper 1973).

4.3 Atrypide Distribution

Cosmopolitan distribution of the early atrypides seems to have continued through the Silurian (Berry and Boucot 1970; Copper 1973), with closely related species occurring in eastern North America and western Europe (United Kingdom, Baltic states). *Protatrypa*, *Gotatrypa*, *Plectatrypa*, *Lissatrypa*, and *Atrypina* are a few of the genera with very wide distribution through the Llandoverian with the Appalachian, Caledonian, Hercynian, Uralian and Cordilleran sea lanes been well connected and open (Copper 1973, Fig. 2). The Llandovery data from ENA incorporated in our research include the following taxa from the Eastern Canada: *Atrypa*, *Dihelictera*, *Gotatrypa*, *Joviatrypa*, *Nalivkinia*, *Protatrypa*, and *Rugosatrypa* which may include some European fauna studied herein due to the mixing and migration patterns. During Wenlock-Ludlow time, later in the Silurian, a few forerunners of the rich Devonian faunas began to appear which include *Eokarpinskia*, *Atrypinella* and *Sibirispira* in the Ural area and *Eospinatrypa* n. gen. in the Eurasia and NA. The Devonian genus *Desquamatia* may have had its roots in the late Llandoverian, and *Atrypa (Atrypa)* became established later in the Silurian (Copper 1973). The Wenlock data in our research was collected from New York, Indiana and Tennessee of ENA which includes the following taxa: *Endrea*, *Atrypa*, *Gotatrypa*, *Oglupes* and *Xanthea*. The Late Silurian data from ENA includes *Atrypa*, *Gotatrypa*, *Oglupes* and *Plectatrypa* from Oklahoma.

Lower Devonian carbonates of the Appalachian province have a limited *Atrypa-Spinatrypa* fauna and our research included *Atrypa-Kyrtatrypa* taxa from ENA localities (New York, Tennessee, Oklahoma, Maryland and West Virginia). During the Middle Devonian, there is an independent development of *Pseudoatrypa* and *Carinatrypa* in Michigan, Ohio and Iowa of ENA (Copper 1973) which does not correlate with the European fauna. However, by the Late Givetian-Early Frasnian, marine connections were reestablished and the presence of *Desquamatia*, *Desquamatia (Independatrypa)*, *Atryparia*, *Atrypa*, *Iowatrypa* was marked both in the Old World and Appalachian Province. *Pseudoatrypa* which originated in the Michigan, New York, Ontario or Ohio during the middle Devonian, extended into the Frasnian of eastern North America, but was absent in western Europe (Copper 1973). Our data includes this taxa *Pseudoatrypa* that persisted from the Middle Devonian (Michigan, New York, Ohio, Missouri and Indiana) to Early Upper Devonian (Iowa) ENA localities. Other taxa included in our analysis from the Middle Devonian localities in ENA include *Atrypa, Desquamatia (Desquamatia)* and *Desquamatia (Independatrypa)*.

4.4 Atrypide Diversity

The Atrypinae were the root stock that appeared in the Silurian (Llandovery) when they diversified and adopted basic characters like the imbricated growth lamellae and frills, with their mode of life less dependent on attachment due to the loss of pedicle muscle. Variatrypinae that appeared during Lower Devonian (Pragian) usually retained the pedicle, developed massive frills, with some shells ranging from reduced growth lamellae to shells with simple tubular rib structure. Members of Variatrypinae dominated the Givetian and Frasnian replacing the member sof Atrypinae subfamily (Copper 2001). However, one could still reasonably derive the Variatrypinae from the Atrypinae subfamily as the two groups were closely related. The geographic endemic center from which the variatrypinids originated and initially dispersed from is unclear, although the oldest described form(s) could help narrow down the particular basin or platform where they originated. Since *Uralospira* is the oldest (Pragian) genus of the Variatrypinae, it is a possible they originated in the vicinity of the Urals, either the eastern European (Russian) platform of eastern Laurussia, or the tropical Siberian platform along the southern margin of the Devonian paleocontinent of Siberia (Day personal communication 2009).

The data analysed in our research from the *Atrypinae* subfamily includes the following taxa: *Atrypa, Gotatrypa, Endrea, Joviatrypa, Protatrypa,* and *Oglupes*. The *Variatrypinae subfamily* includes *Pseudoatrypa, Spinatrypa, Desquamatia (Desquamatia)*, and *Desquamatia (Independatrypa)*. A brief description of each taxa is included below (Copper 1967, 1973, 2001, 2004; Day and Copper 1998).

4.4.1 Atrypinae Subfamily

Atrypa.—Medium sized (23–31 mm width, length similar to width) ovate shaped shell with rounded outline, dorsibiconvex to partly convexoplane, fine rib structure with 8–10 ribs per 5 mm, moderately spaced wave-like growth lamellae (1–3 mm) with their crowding on anterior margins of adult shells, overlapping frills up to 5–8 mm long.

Gotatrypa.—Small to medium, globose shaped shell, ventribiconvex to biconvex to weakly dorsibiconvex, fine to medium ribs intersected with closely spaced wave-like growth lamellae (2–3 mm), with short frills 5 mm long where developed.

Endrea.—Medium to large (21–28 mm width), biconvex-dorsibiconvex, globose, shield-shaped, highly arched tubular imbricate ribs with fine microornament intersecting the ribs, short frills or growth lamellae, commissure weakly to strongly folded.

Joviatrypa.—Similar characteristics as *Gotatrypa* but lacks growth lamellae, if present, only 0.5 mm long.

Protatrypa.—Equally biconvex, subcircular, transverse, or elongate shells commonly without a well developed fold and sulcus. Ornament is similar to *Atrypa*. Relatively flat shell and lacks short, distinct growth lamellae as compared to *Gotatrypa*.

Oglupes.—Large (up to 35 mm), globose and inflated shell, biconvex, moderately coarse ribs with concentric growth lamellae or medium sized to wide frills. Frills relatively wide and prominent as compared to *Gotatrypa* if preserved.

4.4.2 *Variatrypinae Subfamily*

Pseudoatrypa.—Dorsibiconvex-convexiplane, medium-sized, subtriangular to subrectangular, usually strongly uniplicate shells with subtubular-sublamellar ribs and 2–3 mm spaced growth lamellae (no frills or very short projecting frills). Small, ventral beak usually hypercline in maturity, tiny interarea and deltidial plates, foramen subhypothyridid-mesothyridid commonly expanding or enlarging into umbo. *Pseudoatrypa* is most closely related to *Desquamatia*, from which it evolved (Copper 1973). It is distinguishable from that genus by its more closely spaced (more Atrypa-like), growth lamellae by its less tubular ribs, by its tendency to greatly reduce the relative size of the interarea, pedicle opening and deltidial plates in maturity, by its frequent tendency to have a pedicle opening penetrating the umbo, by its much more flattened pedicle valve and strongly arched brachial valve and internally by generally having small dental cavities. *Pseudoatrypa* lacks long frills, which are rarely developed as very short, less than 3 mm extensions. As a whole this genus seems to have lost frilly anchoring devices in favor of flattening the pedicle valve and/or enlarging the foramen through the umbo. *Pseudoatrypa* can be distinguished from *Atrypa* by its interarea, dental cavities and deltidial plates, which are missing in the latter. *Pseudoatrypa* was a soft muddy bottom inhabitant favouring quieter water.

Desquamatia (Desquamatia).—Small size, rounded outline, medium ribs, relatively less widely spaced growth lamellae, rounded hinge line.

Desquamatia (Independatrypa).—Medium to large, dorsibiconvex to convexoplane, subrectangular to shield-shaped atrypid shells with a long hinge line, wide hinge angle, tubular-lamellar ribs, lamellae spaced at regular, relatively large intervals over most of the shell, often with very large multiple frills (largest of any known atrypid group). It can be distinguished from *Desquamatia (Desquamatia)* by its convexity, widely spaced growth lamellae, its generally larger size and its long straight hinge line.

Spinatrypa.—A small-medium sized (largest shells up to 25.5 mm in width, 23 mm in length, 15 mm in thickness); globose dorsibiconvex adult shells with rounded outline; rounded lamellose ribs (7–10 ribs/10 mm) with imbricated spinose lamellae spaced 1.0–2.0 mm on juvenile shells, increasing to 2.0–3.0 mm on larger shells (15 mm) where well preserved; interspaces between ribs up to 1.5 rib width near margins of adults shells; short spines (2–3 mm) arise from concentric lamellae normal to shell surface, rarely preserved on large adult shells, more frequently seen on small juveniles shells; dorsal valve inflated, 1.5–2.0 times as deep as ventral valve.

4.5 Temporal Variation

Phenotypic traits have been studied in the paleontological fossil record both qualitatively and quantitatively in terms of long term and short term geologic intervals. Gradual, directional change was claimed to be rare in the fossil record, with most traits showing little net change except for geologically rapid punctuations associated with speciation (Gould and Elredge 1972). Many aspects of this claim were contentious and specific evolutionary sequences interpreted as gradual by some were seen as representing stasis and/or punctuation by others (Gould and Elredge 1977). Over 250 documented cases of phenotypic traits evolving within fossil lineages were depicted by Hunt (2007) in his study where he observed only 5 % cases indicated directional evolution, 95 % cases involved random walk and stasis patterns with equal chances for each. Gotshall and Lanier (2008) observed an increasing size trend for average body size in brachiopod clades (order and classes) and random unbiased change in average body size at the family level selected from a deep-subtidal, soft substrate habitat during the Cambrian to Devonian (170 Myr) radiation.

Our study herein investigates the average morphological shape trend of atrypide brachiopods over a 64 Myr period from Silurian through Devonian period using different geometric morphometric techniques.

Stasis patterns were observed in Cenozoic mammals by considering the shape of a heuristic time-form evolutionary lattice for longer time scales (Gingerich 2001). Roopnarine (2001) originally interpreted the Mio-Pliocene foraminifers *Globorotalia plesiotumida-tumida* as the punctuated anagenetic transformation of *Globorotalia plesiotumida* to *G. tumida* but using the iterative method of subseries of an original stratophenetic series (by measuring the presence of deviations from statistical randomness as the lineage evolves), he interpreted this as "constrained stasis". Polly (2001) used the Brownian motion model of evolution for reconstructing ancestral nodes versus observed nodes in a fully resolved phylogeny of fossil carnivorans and found that change is constrained in molar areas over longer time intervals. He found that there is change of the sort that one would expect by random selection (Brownian motion) when patterns are measured on a small scale, but that at a very broad scale, (e.g., all mammals) the amount of change was less than if one extrapolated the small scale change to the large scale, suggesting that there is constraint on divergence.

Previous workers (Lieberman et al. 1995) tested the morphological variability of the common brachiopod species lineages from size measurements on the pedicle valves of 401 *Mediospirifer audaculus* and 614 *Athyris spiriferoides* from successive stratigraphic horizons in the Hamilton Group (5 my) of New York. They found morphological overlap within these species between the lowermost and uppermost strata with some variations in the intervening samples of the Hamilton Group (Hamilton EE subunit). Shape variability of the terebratulide brachiopod *Terebratalia transversa* was tested using geometric morphometric technique that gave consistent results in life and death assemblages (Krause 2004). Thus, using geometric morphometric technique for studying our samples from a 64 Myr time interval, was a first time approach because not only that brachiopods were studied first time on a long term scale but also because this test was on atrypide samples, very well preserved and abundant brachiopod group that persisted from Ordovician-Devonian time interval in the Paleozoic rock record (Copper 2001, Fig. 21.1). Morphological shape traits overlapped for both brachial and pedicle valves in atrypides and change was not gradual or directional at all (Figs. 5, 9). However, morphological variability was prominent within atrypide samples (Fig. 5) between successive time intervals corresponding to the EE subunits with little change in mean shape of atrypides within two subfamilies over long time scales (Table 2, Figs. 10–11). Deflection of landmarks along the left posterior margin within Atrypinae subfamily from one time unit to the next suggest some morphological variation within the atrypides during this time which can be attributed to the differing filter-feeding mechanisms and life modes within atrypide taxa of this subfamily. No deflection in landmarks within the Variatrypinae subfamily further suggests stable morphological patterns within these atrypides (Fig. 11) which can be attributed to consistent current flow patterns during feeding of the taxa with similar mode of life within members of this subfamily.

Members of Atrypidina family was dominant in the Silurian and Devonian characterized by ribbed wide shells that developed frills and spines from Wenlock time on with jugal processes nestled between spiralia whorls. Atrypides (Llandeilo-Frasnian) had a complete jugum when they originated but later in the Silurian their jugum split into two jugal processes (Copper 1996). The evolution of these D-shaped spiralia from their planispiral form to dorsomedial forms (with a medially to dorso-medially directed spiralia dominant in Silurian and Devonian) and their jugum should be reflected in the external morphological shape of atrypides through time. Atrypinae and Variatrypinae subfamilies had similar spiralia and jugal processes, thus the overlapping mean morphological shape in our samples (2 subfamilies) reflects less variation in their feeding efficiency. The little valve shape modification observed within these subfamilies could have either been in response of their adapting to changing paleoenvironmental conditions or possibly could have been caused by paleoecological interactions during this time interval. Principal component analysis has shown morphological variation within atrypide samples of these subfamilies (Table 2, Fig. 5). Morphological links and distances between these subfamilies have depicted (Fig. 6, Table 3) that valve shape has slightly fluctuated through time with clear overlap between lowermost and uppermost time intervals suggesting minor shape modification between the two subfamilies. Overall, both brachial and pedicle valves have undergone similar average morphological shape change through time (Table 9) with stasis-like patterns suggested by mean morphological shape trend for both valves (Fig. 10).

4.6   Spatial Variation

Silurian time is a period of marked provincialism for the brachiopod fauna (Boucot and Blodgett 2001), however some early atrypides were cosmopolitan as the seas (Appalachian, Caledonian) was connected during Wenlock-Ludlow. Silurian represents the second Appalachian deformation stage where several disconformities reflect local movements but not really orogenies in the normal sense of that term (Van der Pluijm et al. 1993). Alternating regressive and trangressive phases (Llandoverian–Wenlockian regressive phase (R1) and later Ludlovian–Pridolian second regressive phase (R2)) in the Silurian history (Borque et al. 2000) could be responsible for the significant morphological variation ($p \leq 0.01$) of these well preserved atrypides in Middle Silurian localities (Indiana, Tennessee, New York) (Table 4). The lower and early Middle Devonian (Lockhovian-early Eifelian) time when most fauna were separated by geographic barriers (Rode and Lieberman 2005), and sea level was low, speciation events might have occurred that could be the cause behind significant morphological variation observed in Early Devonian ENA localities (West Virginia, Maryland, Tennessee, Indiana, New York) (Table 5; $p \leq 0.01$). Marked sea-level rise breaching the Transcontinental Arch during the Late Eifelian-Early Givetian time suggests mixing of the fauna from the Old World Realm and the Eastern Americas Realm (Koch and Day 1995; Rode and Lieberman 2005). Late Givetian-Early Frasnian time was marked by decreased provincialism to cosmopolitanism. Very minor significant morphological variation observed in atrypide samples during the Middle Devonian ($p \leq \geq 0.01$) time from different geographic localities (Michigan, New York, Ohio, Indiana, Missouri) in ENA could be attributed to the decreasing provincialism and increasing cosmopolitanism of the brachiopod fauna during the late middle Devonian or could be due to the major transgressive event leading to mixing of the fauna from different geographic realms. Alternatively, this could be due to the presence of closely related species during this time and due to the lack of speciation events (Table 6).

Dendograms illustrated herein depict the morphological (atrypide valve shape) links between different geographic localities from the Eastern North America region during the 3 time intervals sampled. Valve morphological shape in Middle Silurian interval shows a close link between Tennessee and Indiana than with New York (Fig. 6, Tables 7–8). During Early Devonian, samples from Maryland, West Virginia, New York and Tennessee (Appalachian Basin) forms a close cluster with Oklahoma samples (Fig. 7, Tables 7–8). During the Middle Devonian, Missouri samples for brachial valves is more closely linked to those from Michigan than with Indiana, New York and Ohio while for pedicle valves Missouri samples is more closely related to Michigan and New York than with Indiana and Ohio (Fig. 8, Tables 7–8).

Geographic differentiation in morphological shape within atrypide samples differs with respect to brachial and pedicle valves. During the Middle Silurian and Middle Devonian time intervals, brachial valve varies more than the pedicle valve while during Early Devonian, variation in pedicle valves is greater than the brachial valve (Table 10). This might be a clue to the distinct orientation of atrypide taxa living during those time intervals.

4.7 Ecological Causes

4.7.1 Life Orientation and Encruster Preference

Brachial valves were more heavily encrusted than the pedicle valves sampled from ENA localities from Early Silurian through Late Devonian time (Figs. 12, 13) except in the lowermost and uppermost time intervals where they show similar rates of encrustation. Episkeletobionts have been studied with regard to the orientation of brachiopod valves (Rudwick 1962; Spjeldnaes 1984; Hurst 1974; Pitrat and Rogers 1978; Kesling et al. 1980; Bose et al. 2010, 2011; Bose 2013a, b). Most adult atrypide shells live with their brachial valve on top as they lose the pedicle muscle thus giving more exposure time for encrustation. However, some large atrypides anchored to their substrates with their pedicle muscle with the anterior commissure inclined (50–60°) to the sediments (Jone 1982). Encrusting organisms were shown to prefer a specific type of ornament on brachiopods (Schneider and Webb 2004). Encrustation rates, however, did not vary in terms of the ornamentation of the members of the Atrypinae and Variatrypinae subfamilies (Figs. 12–13).

Brachiopod punctae have been suggested to be antifouling and antipredatory defenses. Caeca within the punctae apparently produce chemical deterrents that deter borers (Clarkson 1986). Spicules supporting the caecae may also have an anti-predatory function (Peck 1993). Punctate brachiopods are less frequently encrusted than those that were impunctate (Curry 1983; Thayer 1986). The impunctate spiriferide *Meristella* bore higher rates of encrustation compared to the punctate terebratulides (except *Tropidoleptus*) collected from the Kashong Shale of the Middle Devonian Hamilton Group of New York State (Bordeaux and Brett 1990). Atrypides had an impunctate shell wall thus subject to overall high rates of encrustation (Figs. 12–13; Tables 11–12).

Encrusting organisms are useful as autecological or post-mortem environmental indicators for their hosts, such as whether the host was living at the time of encrustation (Morris and Felton 1993; Lescinsky 1995; Sumrall 2000; Morris and Felton 2003; Schneider 2003, in press; Zhan and Vinn 2007; Rodrigues et al. 2008) or was dead (Thayer 1974; Rodriguez and Gutschick 1975; Anderson and Megivern 1982; Brezinski 1984; Watkins 1981; Gibson 1992). Though it is difficult to decipher whether the atrypide hosts were alive or dead when encrusted, the location of a few encrusters terminating along the anterior margin and along the growth lines suggest they were alive in a few cases. Inhalant feeding currents in atrypides are posterolateral with water entering the gaping shell from either side and exhalant currents are anteromedial (Rudwick-Vogel model) with water leaving the shell through the median deflection antero-centrally (Rudwick 1960, 1970; Mancen˜ido and Gourvennec 2008). The preferential growth of encrusters in these areas of the host are an indicator of the host being alive at the time of encrustation.

Overlapping encrustation rates paralleling the morphological shape overlap between the lowermost and uppermost time units (Lower Clinton and 11. EE subunits) suggest a possibility of ecological stasis driving the morphological conservation observed in the atrypide samples. These atrypide samples were from the EE subunits within the P3 EEU which further emphasizes that 'ecological locking' phenomena could be causing the morphological shape of the atrypides to remain stable through time. However, some deflection of landmarks observed along the left posterior margin of the brachial valves in the atrypides of the Atrypinae subfamily from Early Silurian—Early Devonian (Fig. 10) suggest some variation in morphology of the brachiopods that could have been influenced by the inhalant feeding currents during life of the host.

4.7.2 Encrustation Time and Diversity

Direct determination of the live or dead status of the hosts during encrustation is impossible. Thus, even if a shell is encrusted, there will be no morphological response of the host if it is dead. Spirorbids, cystoporate and trepostome bryozoans, cornulitids, hederellids, auloporid corals and craniid brachiopods were the common encrusters found encrusting the atrypides during this time. Early and Middle Devonian atrypide samples shows higher encrustation rates for brachial valves while only Early Devonian samples for pedicle valves shows higher encrustation rates relative to other time intervals. This is in compliance with the time interval when diverse encrusters evolved.

4.8 Environmental Effect

Atrypides were soft muddy bottom dwellers restricted to shallow water (<100 m deep) tropical shelf environments where environmental conditions (temperature, turbulence, salinity, etc.) were unstable (Fursich and Hurst 1974; Copper 2001).

Silurian atrypides that lived in quiet water (relatively deeper than Devonian atrypides) conditions had their pedicle valve resting on the soft substrate (pedicle opening reduced or absent) and they were smaller in size as the availability of food was less. During later Silurian and Devonian, atrypides living in relatively less quiet water conditions evolved wings and a large pedicle opening to resist the relatively higher energy conditions. They grew larger and thicker taking the advantage of their adequate food availability from shallow water depth. This was relevant from the size and shape data of our Atrypinae and Variatrypinae samples. In general, atrypide brachiopods cannot tolerate turbulent conditions and were all restricted to quiet water energy conditions with fluctuations in temperature, sea level, salinity.

The Plus ~ca change model (Sheldon 1996) predicts a tendency for continuous, gradualistic evolution in the tropical zones and in the deep sea (narrowly fluctuating, relatively stable environments), and more stasis with episodic punctuations in shallow waters and temperate zones (widely fluctuating environments with its application in more physical environmental variables like sea level, substrate, temperature, etc. on a geologic time scale) (Sheldon 1990, 1996, Fig. 2). However, it is well known that trilobites used by Sheldon, include some of the less common genera, i.e., the more rapidly evolving forms. Atrypides sampled for this study belonged to carbonate-siliclastic-mixed lithologic settings with rapid sea level fluctuations (Table 2) within the P3 EEU (Fig. 1). A comparative morphometric shape study on an orthide species from siliciclastic and carbonate setting of the Lower Devonian Hunton Group (Haragan Shale) revealed no morphological shape difference between the samples from these two units (Bose personal observation 2008). Thus, controlling for the lithologic criteria and taking into account the sea level changes during the Silurian and Devonian time intervals, the overall stable mean morphological shape trend of the benthic dwelling atrypides (Fig. 9) on a longer time scale is in accordance with the Plus ca model of Sheldon (1996).

References

Anderson WI, Megivern KD (1982) Epizoans from the Cerro Gordo member of the Lime Creek formation (Upper Devonian). In: Proceedings of the Iowa academy of science, Rockford, Iowa, vol 89, pp 71–80

Berry WBN, Boucot AJ (1970) Correlation of the North American silurian rocks. Geol Soc Am Spec Pap 102:1–289

Bordeaux YL, Brett CE (1990) Substrate specific associations of epibionts on middle Devonian brachiopods: implications for paleoecology. Hist Biol 4:203–220

Bose R, Schneider C, Polly PD, Yacobucci M (2010) Ecological interactions between Rhipidomella (Orthides, brachiopoda) and its endoskeletobionts and predators from the middle devonian dundee formation of Ohio, United states. Palaios 25:196–208

Bose R, Schneider C, Leighton LR, Polly PD (2011) Influence of atrypid morphological shape on Devonian episkeletobiont assemblages from the lower Genshaw formation of the traverse group of Michigan: a geometric morphometric approach. Palaeogeogr Palaeoclimatol Palaeoecol 310:427–441

Bose R (2013a) Devonian paleoenvironments of Ohio, USA. Earth Sciences, Springer p 110. ISBN 978-3-642-34853-2

Bose R, Bartholomew A (2013b) Macroevolution in deep time. Evolutionary Biology, Springer p 75. ISBN 978-1-4614-6475-4

Boucot AJ, Blodgett RB (2001) Silurian-Devonian biogeography. In: Brunton HC, Cocks RM, Long SL (eds) Brachiopods past and present the natural history museum. Taylor and Francis Publishers, London, pp 335–344

Bourque PA, Malo M, Kirkwood D (2000) Paleogeography and tectono-sedimentary history at the margin of Laurentia during Silurian to earliest Devonian time: the Gaspé Belt, Québec. GSA Bulletin 112:4–20

Brezinski DK (1984) Upper Mississippian epizoans and hosts from southwestern Pennsylvania. Proc PA Acad Sci 58:223–226

Clarkson ENK (1986) Invertebrate paleontology and evolution, 2nd edn. Allen and Unwin, London, p 382

Copper P (1967) Pedicle morphology in Devonian atrypid brachiopods. J Paleontol 41:1166–1175

Copper P (1973) New Siluro-Devonian atrypoid brachiopods. J Paleontol 47:484–500

Copper P (2001) Radiations and extinctions of atrypide brachiopods: ordovician—Devonian. In: Brunton CHC, Cocks LRM, Long SL (eds) Brachiopods past and present. Natural History Museum, London, pp 201–211

Copper P (2004) Silurian (Late Llandovery-Ludlow) Atrypid brachiopods from Gotland, Sweden, and the Welsh Borderlands, the Great Britain. National Research Council of Canada Research Press, Ottawa

Copper P, Gourvennec R (1996) Evolution of the spire-bearing brachiopods (Ordovician-Jurassic). In: Proceedings of the third international brachiopod congress, pp 81–88

Curry GB (1983) Brachiopod caeca—a respiratory role? Lethaia 16:311–312

Day J (1998) Distribution of latest Givetian-Frasnian Atrypida (Brachiopoda) in central and western North America. Acta Palaeontol Pol 43:205–240

Eldredge N, Gould SJ (1972) Punctuated equilibria: an alternative to phyletic gradualism. In: Schopf TJM (ed) Models in paleobiology. Freeman, San Francisco, pp 82–115

Fursich FT, Hurst JM (1974) Environmental factors determining the distribution of brachiopods. Paleontology 17:879–900

Gibson MA (1992) Some epibiont-host and epibiont-epibiont relationships from the Birdsong shale member of the lower Devonian Ross formation (west-central Tennessee, USA). Hist Biol 6:113–132

Gingerich PD (2001) Rates of evolution on the time scale of the evolutionary process. Genetica 112–113:127–144

Gould SJ, Eldredge N (1977) Punctuated equilibria: the tempo and mode of evolution reconsidered. Paleobiology 3:115–151

Hunt G (2007) The relative importance of directional change, random walks, and stasis in the evolution of fossil lineages. Proc Natl Acad Sci 104:18404–18408

Jones B (1982) Paleobiology of the upper Silurian Brachiopod Atrypoidea. J Paleontol 56:912–923

Kesling RV, Hoare RD, Sparks DK (1980) Epizoans of the middle Devonian brachiopod *Paraspirifer bownockeri*: their relationships to one another and to their host. J Paleontol 54:1141–1154

Koch W, Day J (1995) Late Eifelian-early Givetian (Middle Devonian) brachiopod paleobiogeography of eastern and central North America. Brachiopods: Proceedings of the third international Brachiopod congress, vol 3, pp 135–143

Krause RA (2004) An assessment of morphological fidelity in the sub-fossil record of a Terebratulide Brachiopod. Palaios 19:460–476

Lescinsky HL (1995) The life orientation of concavo-convex brachiopods: overturning the paradigm. Paleobiology 21:520–551

Lieberman BS, Brett CE, Elredge N (1995) A study of stasis in two species lineages from the Middle Devonian of New York State. Paleobiology 21:15–27

Lowenstam HA (1957) Niagaran reefs in the great lakes area. Geol Soc Am Mem 67:215–248

Mancenido MO, Gourvennec R (2008) A reappraisal of feeding current systems inferred for spire-bearing brachiopods. Earth Environ Sci Trans R Soc Edinb 98:345–356

Morris RW, Felton SH (1993) Symbiotic association of crinoids, platyceratid gastropods, and cornulites in the Upper Ordovician (Cincinnatian) of the Cincinnati Ohio region. Palaios 8:465–476

Morris RW, Felton SH (2003) Paleoecologic associations and secondary tiering of cornulites on crinoids and bivalves in the Upper Ordovician (Cincinnatian) of southwestern Ohio, southeastern Indiana, and northern Kentucky. Palaios 18:546–558

Novack-Gottshall PM, Michael AL (2008) Scale-dependence of Cope's rule in body size evolution of Paleozoic brachiopods. Proc Natl Acad Sci 105:5430–5434

Peck LS (1993) The tissues of articulate brachiopods and their value to predators. Phil Trans Biol Sci 339:17–32

Pitrat CW, Rogers FS (1978) Spinocyrtia and its epizoans in the traverse group (Devonian) of Michigan. J Paleontol 52:1315–1324

Polly PD (2001) Paleontology and the comparative method: ancestral node reconstructions versus observed node values. Am Nat 157:596–609

Rode AL, Lieberman BS (2005) Integrating evolution and biogeography: a case study involving Devonian crustaceans. J Paleontol 79:267–276

Rodrigues SC, Simoes MG, Kowalewski M, Petti MAV, Nonato EF, Martinez S, Rio CDJ (2008) Biotic interaction between spionid polychaetes and bouchardiid brachiopods: paleoecological, taphonomic and evolutionary implications. Acta Palaeontol Pol 53:657–668

Rodriguez J, Gutschick RC (1975) Epibiontic relationships on a late Devonian algal bank. J Paleontol 49:1112–1120

Roopnarine PD (2001) The description and classification of evolutionary mode: a computational approach. Paleobiology 27:446–465

Rudwick MJS (1960) The feeding mechanisms of spire-bearing fossil brachiopods. Geol Mag 97:369–383

Rudwick MJS (1962) Filter-feeding mechanisms in some brachiopods from New Zealand. J Linn Soc Lond Zool 44:592–615

Rudwick MJS (1970) Living and fossil brachiopods. Hutchinson University Library Press, London, p 199

Schneider CL (2003) Hitchhiking on Pennsylvanian echinoids: epibionts on Archaeocidaris. Palaios 18:435–444

Schneider CL, Webb A (2004) Where have all the encrusters gone? encrusting organisms on Devonian versus Mississippian brachiopods. Geol Soc Am Abstr 36:111

Sheldon PR (1990) Shaking up evolutionary patterns. Nature 345:772

Sheldon PR (1996) Plus qa change—a model for stasis and evolution in different environments. Palaeogeogr Palaeoclimatol Palaeoecol 127(209):227

Spjeldnaes N (1984) Epifauna as a tool in autecological analysis of Silurian brachiopods: special papers in Paleontology, vol 32, pp 225–235

Sumrall CD (2000) The biological implications of an edrioasteroid attached to a pleurocystitid rhombiferan. J Paleontol 74:67–71

Thayer CW (1974) Substrate specificity of Devonian epizoa. J Paleontol 48:881–894

Thayer CW (1986) Respiration and the function of brachiopod punctae. Lethaia 19:23–31

Van der Pluijm B (1993) Paleogeography, accretionary history, and tectonic scenario; a working hypothesis for the Ordovician and Silurian evolution of the northern Appalachians. special paper. Geol Soc Am 275:27–40

Watkins R (1981) Epizoan ecology of the type Ludlow series (Upper Silurian), England. J Paleontol 55:29–32

Zhan R, Vinn O (2007) Cornulitid epibionts on brachiopod shells from the late Ordovician (middle Ashgill) of East China. Est J Earth Sci 56(2):101–108

Chapter 5
Conclusion

Morphological shape is relatively stable in Silurian–Devonian atrypide members within the P3 EEU. Results from several geometric morphometric techniques (including Procrustes analysis and Thin-plate spline) confirm a high degree of morphological variability within samples from each time unit with little change in mean shape between atrypide individuals from Atrypinae and Variatrypinae sub-families. Geographic differentiation in morphological shape within atrypide samples appears to be greater than temporal variation.

Morphological overlap was clearly depicted between lowermost and uppermost time intervals within atrypide individuals. Morphological shape does not vary much between the two subfamilies which further acts as a strong evidence for the similarity in lophophore and the brachidium shape within these extinct taxa. Lowermost and uppermost time intervals show similar encrustation rates with high rates of encrustation in Lower-Middle Devonian. Perhaps, ecological stasis could be driving the morphological shape change patterns in the long time scale interval resulting in stable morphological shape within the P3 EEU. Brachial valve shows higher rates of encrustation as compared to pedicle valves in atrypides which is possibly due to the pedicle valve resting on the substrate and the brachial valve oriented upside in absence of pedicle muscles in adult shells. Overall, in the 64 myr time scale within the P3 EEU, atrypide morphological shape varies insignificantly in widely fluctuating environments. This suggests stasis-like patterns with moderate geographic variations that could further throw light on the distribution of atrypide taxa and their likely speciation events that occurred during those time intervals.

Finally, question arises whether these atrypides that show good evidence for stasis can be assumed to "represent" all other taxa. Since atrypaceans are one of the "dominant" taxa numerically in the communities from which my samples were drawn, this may probably be the case (Boucot et al. 2012).

One exceptional study by De and Bose (2009) speculated hypotheses to probe an evolutionary pathway in invertebrates at the genome level. Perhaps, future studies in exploring the evolutionary pathway of marine invertebrate lineages using natural experiments, could throw new insights into their ancestral origin.

R. Bose, *Palaeobiology of Middle Paleozoic Marine Brachiopods*, SpringerBriefs in Earth Sciences, DOI: 10.1007/978-3-319-00194-4_5, © The Author(s) 2013

References

De A, Bose R (2009) Can molecular biology and bioinformatics be used to probe an evolutionary pathway? Proc Nat Acad Sci USA 106:52

Boucot AJ, Blodgett RB, Rohr DM (2012) Brachiopoda (Atrypidae), from upper Silurian strata of the Alexander terrane, northeast Chichagof Island, Alaska. Bull Geosci 87:261–267

About the Author

Dr. Rituparna Bose obtained her Ph.D. from Indiana University, Bloomington. For her outstanding Ph.D. dissertation she was rewarded with a Springer Theses Prize. She is currently an adjunct Assistant Professor at the City University of New York and has been interviewed as an expert in the field of biodiversity by the Times of India (leading news daily in her home country). She has won numerous awards in her career including the nationally recognized Theodore Roosevelt Memorial Grant awarded by the American Museum of Natural History.

She has been recently invited to serve as an Editor for *Acta Paleontologica Sinica* by the Chinese Academy of Sciences. She is the Associate Editor-in-Chief for the International Journal of Environmental Protection and the Associate Editor for the *Journal of Geography and Geology* at the Canadian Center of Science. She also serves on the editorial board of some of the most prestigious journals in geology including *Historical Biology: An International Journal of Palaeobiology* (Taylor and Francis), *Bulletins of American Paleontology* (Paleontological Research Institute, Cornell University), *Springer Plus* (Earth and Environmental Science), and *Geological Journal* (Wiley).

R. Bose, *Palaeobiology of Middle Paleozoic Marine Brachiopods*,
SpringerBriefs in Earth Sciences, DOI: 10.1007/978-3-319-00194-4,

Curriculum Vitae

Dr. Rituparna Bose
Adjunct Assistant Professor
The City University of New York

Research Interests

The prerequisite to developing effective strategies for conserving biodiversity is a profound understanding of the taxonomy and phylogeny of all life forms. It is especially important to appreciate the significance of such studies in extinct organisms; especially in organisms that were abundant in a certain geologic era, but have subsequently dwindled or become extinct. Such studies should help to understand extinction, accurately gauge the underlying causes behind loss of biodiversity and make predictions about future distribution of biodiversity. I have developed models during my graduate career to trace loss of biodiversity, and what should also serve as a starting point for conservation. My research interests encompass paleontology, geobiology, micropaleontology, sedimentary geochemistry and paleoceanography; and have implications for the petroleum industry as research could potentially uncover new oil-rich locations.

R. Bose, *Palaeobiology of Middle Paleozoic Marine Brachiopods*,
SpringerBriefs in Earth Sciences, DOI: 10.1007/978-3-319-00194-4,
© The Author(s) 2013

Education

Doctor of Philosophy (PhD), Geological Sciences (Major: Geobiology & Paleontology; Minor: Hydrogeology), 2011
Department of Geological Sciences, Indiana University, Bloomington, IN
Advisor: Dr. P. David Polly
Dissertation: Morphometric evolution of Paleozoic brachiopods: a geometric morphometric approach

Master of Science (MS), Geology, 2006
Bowling Green State University, Bowling Green, OH
Advisor: Dr. Margaret Yacobucci
Thesis: Epibionts on brachiopods from the Devonian Dundee formation of Ohio

Master of Science (MSc), Applied Geology, 2003
Jadavpur University, Calcutta, India, 1st class Honors; ranked second in University

Bachelor of Science (BSc), *Major*: Geology (Honors); *Minors*: Maths. & Chem., 2001
University of Calcutta, India, 1st class Honors; ranked second in college

Awards and Honors

Best thesis award in Springer Theses series - 'the best of the best', Recognizing outstanding PhD research, 2012 (€ 500)

Dissertation Year Research Fellowship, Indiana University College of Arts and Sciences, 2010–2011, ($18,000)

Theodore Roosevelt Memorial Grant, American Museum of Natural History, 2009, ($2000)

Schuchert and Dunbar Grant, Yale Peabody Museum of Natural History, 2009, ($1000)

Galloway-Horowitz Summer Research Grant Recipient, Department of Geological Sciences, Indiana University, Bloomington, 2008 and 2009 ($3000 and $1500)

North American Paleontology Convention (NAPC) Travel Award Recipient, funded by BP Global Energy Group, 2009, ($250)

Graduate Fellowship, Indiana University, 2006, ($1000)

Graduate Research Scholarship, Jadavpur University, India, 2003, (Rs. 1000)

Research Experience

Museum research: National Museum of Natural History, Yale Peabody Museum, National Museum of Natural History, American Museum of Natural History, Michigan Museum of Paleontology at Ann Arbor, New York State Museum (Albany and New Paltz), Indiana University Paleontology Collections, 2006–present

Field Geology, Sedimentology, Stratigraphy, Paleontology, etc., 1998–present

Science writing: Extensive experience in reading, writing and publishing in the field of Science, 2003–present

Peer-review: Reviewing, critiquing, and editing other research papers, 2003–present

Stereomicroscopy, Geological specimens, 1998–present

Scanning electron microscopy, Biological study, 2006–present

3-D laser scanner imaging, Morphometric study, 2006–present

Environmental and geological modeling (ArcGIS), 2006–present

Biostatistics (use of statistical softwares R, PAST, TPS, etc.), 2007–present

Photography and digitizing samples using microscope and SPOT software

Analytical skills in Geochemistry, Hydrogeology, 2006–2008

Teaching Experience

Summer Principal Instructor, 2012
    CUNY Urban Scholars Program at the City College of New York
    Courses taught: Living Environment, Weather and Climate

Adjunct Assistant Professor, 2011–present
    Department of Earth and Atmospheric Sciences
    The City College of New York, CUNY, New York, NY
    http://www.adm.ccny.cuny.edu/v2/directory/dirfind.cfm?urltarget=Earth%20
    and%20Atmospheric%20Science
    Course sections taught: Global Natural Hazards using GIS applications
    (EAS328), Earth System Science (EAS106)

Adjunct Assistant Professor, 2011–present
    Department of Biological Sciences and Geology
    Queensborough Community College, CUNY, New York, NY
    http://www.qcc.cuny.edu/directory/FacultyDetail.aspx?personID=2874
    Course sections taught: Physical Geology (GE101), Historical Geology
    (GE102), Earth Science (GE125)

Guest lecturer, 2010
Department of Geological Sciences
Indiana University, Bloomington, IN
Course section taught: Evolutionary Ecology (GEOL 690)

Principal Instructor, 2009 and 2010
Department of Geological Sciences
Indiana University, Bloomington, IN
Course taught: Earth, Our Habitable Planet (GEOL105)

Principal Instructor, SU 2007
Upward bound Program
Smith Research Center, School of Education
Indiana University, Bloomington, IN
Course taught: Geology

Associate Instructor/Teaching Assistant, 2006–2010
Department of Geological Sciences
Indiana University, Bloomington, IN
Courses taught: General Geology, Physical Geology, Historical Geology,
Theory of the Earth (Honors topic), Earth Science - Materials and Processes,
Dinosaurs and their relatives, Evolution of the Earth

Teaching Assistant, 2004–2006
Department of Geology
Bowling Green State University, Bowling Green, OH
Courses taught: Historical Geology, Dinosaurs

List of Publications: Books/Scientific Monographs/Book Chapters

Bose, R., 2012. Biodiversity and evolutionary ecology of extinct organisms. *Springer Verlag book series*. 214 p. ISBN 978-3-642-31720-0.

Bose, R., 2013. Devonian paleoenvironments of Ohio, USA. *Springer Verlag book series*. 57 p. ISBN 978-3-642-34854-9.

Bose, R., 2012. An introduction to Geobiology of marine organisms. Manuscript under review in *Springer Briefs*.

Bose, R., 2013. Macroevolution in deep time. *Springer verlag book series*. 59 p. ISBN 978-1-4614-6475-4.

Manuscripts Accepted/in Press/Published

Bose, R., 2012. A new morphometric model in distinguishing two closely related extinct brachiopod species. *Historical Biology* 24: 1–10.

Bose, R., 2012. Quantitative analysis strengthens qualitative assessment: A case study of Devonian brachiopod species. *Paläontologische Zeitschrift, Scientific Contributions to Palaeontology*: 1–8.

Bose, R., De, A., Mazumdar, S., Sen, G., & Mukherjee, A. D., 2012. Comparative study of the physico-chemical parameters of the coastal waters in river Matla and Saptamukhi: impacts of coastal water coastal pollution. *Journal of Water Chemistry and Technology* 34(5):1–10.

Bose, R., and De, A., 2012. Impacts of coastal water pollution in Matla and Saptamukhi rivers of the Bengal delta. *Geochemistry International* 50 (10): 1–10.

Bose, R., Schneider, C., Leighton, L. R., and Polly, P. D., 2011. Influence of atrypid morphological shape on Devonian episkeletobiont assemblages from the lower Genshaw Formation of the Traverse Group of Michigan: a geometric morphometric approach. *Paleogeography, Paleoecology and Paleoclimatology* 310: 427–441.

De, A., and **Bose, R.,** 2011. Smart Science Investment in India. *Reply to* India's Vision: From Scientific Pipsqueak to Powerhouse. *Science E-letters*. http://www.sciencemag.org/content/330/6000/23.1/reply

Bose, R., 2011. Morphometric evolution of Paleozoic brachiopods - the effects of environment and ecological interactions on shell morphology, PhD thesis, 199 p. http://upload.etdadmin.com/etdadmin/files/102/125119_pdf_117050_A058E4BC-2756-11E1-9780-0D3CEF8616FA.pdf

Bose, R., Schneider, C., Polly, P. D., and Yacobucci, M., 2010. Ecological interactions between *Rhipidomella* (Orthides, brachiopoda) and its endoskeletobionts and predators from the Middle Devonian Dundee Formation of Ohio, United states. *Palaios* 25:196–208.

De, A., Mozumdar, S., and **Bose, R**. 2010. Taste receptors, chemical kinetics and equilibrium. *The Scepticalchymist, The Nature Chemistry Blog*. http://blogs.nature.com/thescepticalchymist/2010/06/

De, A., and **Bose, R.,** 2009. Can molecular biology and bioinformatics be used to probe an evolutionary pathway? *Proceedings of the National Academy of Sciences* 106:E141 http://www.pnas.org/content/106/52/E141.full.pdf+html

Bose, R., 2006. Epibionts on brachiopods from the Devonian Dundee formation of Ohio, MS thesis, Bowling Green State University, 84 p. http://www.ohiolink.edu/etd/view.cgi?acc_num=bgsu1155059386

Manuscripts Under Review

Bose, R., and De, A., 2012. Quantitative evaluation reveals taxonomic over-splitting in extinct brachiopods from deep time: implications in conserving biodiversity. Manuscript under review in *Plos One*. 4

Bose, R., and De, A., 2012. Arsenic contamination - unavoidable natural phenomenon or an anthropogenic crisis. Manuscript under review in *Environmental Earth Sciences*.

Bose, R., Leighton, L. R., Day, J., and Polly, P. D., 2012. Use of shell shape as proxy in testing the taxonomy and phylogeny of Paleozoic atrypids in eastern North America. Manuscript under review in *Journal of Paleontology*.

Manuscripts in Preparation

Bose, R., and Polly, P., 2012. Body shape evolution in Paleozoic brachiopod species. Manuscript under preparation for submission to *Paleobiology*.

Bose, R., 2012. Morphological trends in brachiopod species across middle Devonian stratigraphic units (Traverse and Hamilton Group) of Eastern North America. Manuscript under preparation for submission to *Swiss Journal of Paleontology*.

Bose, R., and Polly, P. D., 2012. Morphological evolution in *Pseudoatrypa cf. lineata* from the middle Devonian Traverse Group of Michigan. Manuscript in preparation for submission to *Journal of Evolutionary Biology*.

Presentations at Regional and National Conferences

Bose, R., Polly, P. D., 2012. Evolution on a temporal scale: A case study of a Devonian brachiopod species in the Traverse group of north eastern Michigan, USA. *Geological Society of America Abstracts with Programs*, xx, p. xxx.

Bose, R., De, A., Polly, P. D., 2012. Quantitative evaluation reveals taxonomic over-splitting in extinct brachiopods from deep time: implications in conserving biodiversity. *The Society For The Preservation of Natural History Collections*, 27 (48).

Bose, R., Leighton, L. R., and Polly, P. D., 2010. A geometric morphometric study of Silurian and Devonian brachiopod subfamilies of the Atrypida order (Atrypinae, Variatrypinae and Spinatrypinae) from Eastern North America, *Geological Society of America Abstracts with Programs*, 42, p. 142. http://gsa.confex.com/gsa/2010AM/finalprogram/abstract_182095.htm

Bose, R., and Schneider, C., 2010. Episkeletobionts on *Pseudoatrypa* (brachiopoda) from the lower Genshaw Formation of the Middle Devonian Traverse Group of Michigan, USA, *Geological Society of America Abstracts with Programs,* 42: 98. http://gsa.confex.com/gsa/2010AM/finalprogram/abstract_181536.htm

Bose, R., Schneider, C., Polly, P. D., and Yacobucci, M., 2009. Synecological interactions of the brachiopod fauna in the carbonate environment of the Middle Devonian Dundee Formation of Ohio, USA. Presented at *9th North American Paleontological Convention.*

Bose, R., and Polly, P. D., 2009. Evolutionary change and geographic variation in Silurian-Devonian *Atrypa reticularis:* A geometric morphometric approach. Presented at *9th North American Paleontological Convention.*

Bose, R., and Polly, P. D., 2009. Evolution in Paleozoic brachiopod morphology: a geometric morphometric approach. Presented at *CGC 8th Annual Conference* and *11th Women in Science Research Conference,* Indiana University.

Bose, R., 2008. Mandible and premolar shape variation in mammalian carnivores and their association with diet. *CGC 7th Annual Conference Abstracts,* Indiana University.

Participation in *Geological Society of America conference,* 2007.

Bose, R., 2007. Epibionts on brachiopods from the Devonian Dundee Formation of Ohio. *Dept. of Geol. Sc. day 6th Annual Conference Abstracts,* Indiana University, p 17. http://www.indiana.edu/~geosci/dept/dogsrd/2007dogsdayprogram.pdf

Bose, R., Yacobucci, M. and Wright, C., 2006. Epibionts on the brachiopods from the Devonian Dundee Formation of Ohio. *Geological Society of America Abstracts with Programs,* 38, p 14. http://gsa.confex.com/gsa/2006nc/finalprogram/abstract_103661.htm

Professional Activities

Book Proposals Under Review

Special Issue in Philosophical Transactions of the Royal Society B—Genetics of extinct organisms

Special Issue in Natural Science—Earth and Environmental Sciences

Special Issue in American Journal of Climate Change—Hydrology and Global Environmental Change

Book in Springer Topics in Geobiology series—Evolutionary Ecology of Brachiopods

Book in Springer Coastal Research Library series—Integrated Coastal Zone Management and Coastal and marine biology

Book in InTech Open Science—Application of Geosciences in other Applied Sciences

Editorial Advisory Board Members

Historical Biology: An International Journal of Palaeobiology (Taylor and Francis)
Bulletins of American Paleontology, Cornell University
Palaeontographica Americana, Cornell University

Peer Review for Journals

AGU journals:
Journal of Geophysical Research: Oceans (American Geophysical Union)
Water Resources Research (American Geophysical Union)
Geology (Geological Society of America)

University and Museum publishers
Bulletin of the Museum of Comparative Zoology, Harvard University
Bulletin of the American Museum of Natural History, American Museum of Natural History
Bulletins of American Paleontology, Cornell University
Palaeontographica Americana, Cornell University
Bulletin of the Peabody Museum of Natural History, Yale University

Royal Society journals
Proceedings of the Royal Society A: Mathematical, Physical and Engineering Sciences
Philosophical Transactions of the Royal Society B: Biological Sciences

Elsevier journals
Palaios (Elsevier)
Paleogeography, Paleoecology and Paleoclimatology (Elsevier)

Springer journals
Environmental Earth Sciences, Springer
Environmental Geochemistry and Health, Springer
Water Quality, Exposure and Health, Springer
Paleontological Journal, Springer
Swiss Journal of Palaeontology, Springer
Journal of Applied Water Science, Springer Open

International Journal of Recycling of Organic Waste in Agriculture, Springer Open
International Aquatic Research, Springer Open

Other publishers
Ecology and Evolution (Wiley)
Historical Biology: An International Journal of Palaeobiology (Taylor and Francis)
Journal of Coastal Research, Coastal Education and Research Foundation (CERF)
Journal of Geography and Geology, Canadian Center of Science and Education
Acta Palaeontologica Polonica, Polish Academy of Sciences
American Malacological Bulletin, The American Malacological Society
BIOS: a quarterly journal of Biology, Beta Beta Beta Biological Honor Society
Journal of Trace Element Analysis, Columbia International Publishers (CIP)
American Journal of Climate Change, Scientific Research Publishing
African Invertebrates, Council of the KwaZulu-Natal Museum
Journal of East African Natural History, Nature Kenya/East African Natural
History Society
American Journal Experts
PlosOne, Public Library of Science

In the News

Short story on **"Prehistoric finds"**, Mississipian spiriferids, a new finding from
Mooresville locality, Indiana published in **MD times newspaper**, 2009 (8th
April)   (http://www.reporter-times.com/stories/2009/04/08/news.nw-206161.sto);
Participation in **Friday Zone Episode** 'Indiana Fossils', 2009

Outreach and Other Activities

Prepared high school kids for National level Science Olympiad Competitions,
2008–2010
Prepared a project for Brownie Math and Science Major junior students, 2009
Supervised two undergraduate students in paleontological research in graduate stu-
dent career, 2008–2011
GSA Short course in "Geoscience teaching at high school level", GSA, Denver,
CO, 2007

Society Memberships

Geological Society of America, Paleontological Society, The Society for the
Preservation of Natural History Collections